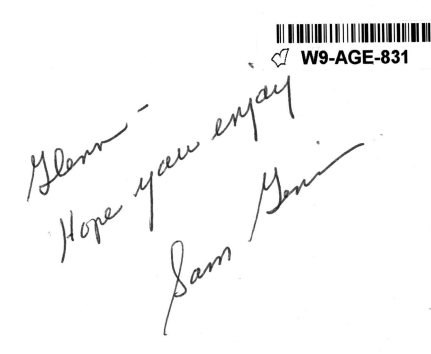

Glenn –
Hope you enjoy

Sam Geni

Anytime, Anywhere
Entrepreneurship and the Creation of a Wireless World

Wireless entrepreneurs are transforming the way people live and work around the globe. In the process they have created some of the fastest-growing companies on the planet. This book candidly tells the story of the birth and explosion of cellular and wireless communications by focusing on one of the industry's pioneers, Sam Ginn. As deregulation and privatization swept the globe, Ginn and his team at AirTouch Communications fought for and won licenses on several continents. They built an amazingly successful business using strategic partnerships and joint ventures. In the process, they demonstrated a new model for global entrepreneurship in a high-tech, information-based economy. The combination of AirTouch with Vodafone in 1999, and of Vodafone with Mannesmann in Europe in 2000, has created the largest wireless business in the world. Vodafone also formed a joint venture with Bell Atlantic to create Verizon Wireless in 2000, the largest wireless company in the United States.

Louis Galambos is Professor of History at Johns Hopkins University in Baltimore, Maryland, and editor of *The Papers of Dwight David Eisenhower*. He has written numerous books and articles on entrepreneurship, innovation, and regulation. He has co-authored books on pharmaceuticals, public policy, and telecommunications, including *Networks of Innovation* (Cambridge, 1995), *The Rise of the Corporate Commonwealth* (1990), and *The Fall of the Bell System* (Cambridge, 1986). He is president of the Business History Group.

Eric John Abrahamson is Principal Historian with The Prologue Group. He has written about telecommunications, banking, legal education, and regulation in California, including a 1994 study of the Pacific Telesis Group. His research at Johns Hopkins University has dealt with mobile telephony as well as the birth and management of customer service in the United States and Europe in the early twentieth century.

Dedicated
to our patient, wise, supportive, good-humored families

Anytime, Anywhere

Entrepreneurship and the
Creation of a Wireless World

Louis Galambos

Johns Hopkins University

Eric John Abrahamson

The Prologue Group

CAMBRIDGE
UNIVERSITY PRESS

PUBLISHED BY THE PRESS SYNDICATE OF THE UNIVERSITY OF CAMBRIDGE
The Pitt Building, Trumpington Street, Cambridge, United Kingdom

CAMBRIDGE UNIVERSITY PRESS
The Edinburgh Building, Cambridge CB2 2RU, UK
40 West 20th Street, New York, NY 10011-4211, USA
477 Williamstown Road, Port Melbourne, VIC 3207, Australia
Ruiz de Alarcón 13, 28014 Madrid, Spain
Dock House, The Waterfront, Cape Town 8001, South Africa

http://www.cambridge.org

First published 2002

Printed in the United States of America

Typeface Sabon 10/13 pt.   *System* AMS-TEX   [FH]

A catalog record for this book is available from the British Library.

Library of Congress Cataloging in Publication data available

ISBN 0 521 81616 5 hardback

Contents

Contents

Acknowledgments

Any history that covers a major industry in the global economy depends on the favors of many individuals and institutions scattered across various continents. We conducted more than a hundred interviews for this work, talking to current and former employees and executives from AT&T, Pacific Telesis Group, AirTouch, Communications Industries, Mannesmann, Europolitan, and Vodafone Group, as well as staff and commissioners at the California Public Utilities Commission. We thank all of the people who took time to talk to us and provide information and insights that were invaluable in developing our narrative. In alphabetical order, they are Robert Barada, Michael Boskin, Charles Brown, John Bryson, Mike Caldwell, C. Lee Cox, Robert Dalenberg, Amy Damianakes, Carl Danner, Virginia Dwyer, Patricia Eckert, Thomas Ehler, William Ellinghaus, F. Craig Farrill, Chris Gent, Ann Ginn, Myra Ginn, Sam Ginn, Sy Graff, Annelie Green, Mohan Gyani, Mark Hickey, Brenda Hooks, John Hulse, Thomas Isakson, Dwight Jasmann, Craig Jorgens, Keith Kaczmarek, William Keever, Terry Kramer, Barry Lewis, Craig McCaw, James McCraney, Erhart Meixner, Michael Miron, Jan Neels, Clayton Niles, Paul Popenoe, Phil Quigley, Kathleen Reinhart, Barbara Riker, Arthur Rock, Arun Sarin, George Schmitt, George Shultz, Cindy Silva, Don Sledge, Robert Smelik, Tommy Sundstrom, Vern Tyerman, April Walden, and Erik Young. A special acknowledgment goes to Erhart Meixner, who twice exceeded the reasonable requests of authors and went out of his way to assist our cause.

Many people also helped by reading some or all of the manuscript and offering suggestions. We thank for their time and their ideas all of the following: Reed Abrahamson, Glenn Bugos, Mike Caldwell, Amy Damianakes,

Acknowledgments

Patricia Eckert, Melanie Fannin, Sam Ginn, Erhart Meixner, Jan Neels, Clayton Niles, Arun Sarin, George Schmitt, and Eric Schuster. Thanks also to the anonymous readers at Cambridge University Press who provided a number of constructive ideas.

Our editor at Cambridge, Frank Smith, performed brilliantly, as usual, and we are also deeply indebted to Cathy Felgar and Barbara Chin, both of whom helped shepherd our book through the Press and into the hands of our readers. We are grateful as well to Helen Rees, our agent, for all of her efforts on our behalf.

Some of the primary material used to tell this story was developed on a project for Pacific Telesis Group in the early 1990s, a study that focused on the breakup of the Bell System in the context of California regulatory politics. Our thanks to SBC Corporation and its chairman Ed Whitacre for allowing us to draw on that study as we did the research for the present manuscript. Marjorie Wilkins was a co-author on the project for Pacific Telesis, and her advice and encouragement on this present undertaking were much appreciated. Thanks also to her partner, Joyce Vollmer, who helped brainstorm ideas when we were just getting underway.

We received absolutely essential support from AirTouch, where Mark Hickey championed and guided this project over a number of years. Corporate Librarian Deena Karadesh helped us track down industry materials and Archivist Tracey Panek ensured that the record of AirTouch's development was kept intact and made available to us. She also saw to it that we were able to use the interviews she had conducted. All of our visits to AirTouch were handled by Gerry Garber, the organizational whiz in Sam Ginn's office, who assisted us in gaining access to the people and materials we needed, when we needed them.

Sam Ginn gave us all the help we needed while leaving in our hands and minds all of the important interpretive questions raised by this study. He took part in numerous interviews and gave us uncensored access to his personal papers. He made it possible for us to interview most of the other busy entrepreneurs, executives, and officials who played important roles in the history of the wireless industry. Like his mentor Charlie Brown, Sam Ginn's only demand on the authors was that we write an accurate history, warts and all, of these important and exciting events.

At Johns Hopkins University, we had the enthusiastic support of Elizabeth Kafig, who kept us from making numerous errors as the project evolved. She was ably supported by Mary Butler Davies, Jill Friedman, and Sharon Widomski, the master organizer of the Department of History at Hopkins. Two departmental chairs, professors John Russell-Wood and

Acknowledgments

Gaby Spiegel, were helpful throughout our labors, as were various persons at the Milton S. Eisenhower Library. We also thank the numerous participants in the seminars on innovation conducted at Hopkins by the Institute for Applied Economics and the Study of Business Enterprise; they helped us improve our understanding of entrepreneurship in its global setting.

In the American heartland, the staff of the Rapid City (South Dakota) Public Library – especially interlibrary loan ace Jason Walker – helped erase the liabilities of distance while Craig Chapman superbly transcribed all of our interviews and generated stacks of background research. Ernie Grafe created tables and charts, critiqued the manuscript, and prepared the index. Zachary Abrahamson helped compile research and proofed pages. Neil Chamberlain good-naturedly corrected many technical gaffes and shared a fascination for the wireless industry. We also thank Motorola for providing access to its corporate archives. Further west, Eric received refuge and square meals from Joanne and Tony Smith, who enabled him to have easy access to the libraries at Stanford University and archives in San Francisco.

Both of us are indebted to our colleagues at the Business History Group and The Prologue Group. William and Ruth Anne Becker provided encouragement and made sure the bills got paid. Glenn Bugos's insights were always to the point and constructive; his sense of humor kept things in perspective.

Our families, to whom we have dedicated this book, put up with incessant phone calls, e-mail messages, trips, complaints about impossible co-authors, and above all, the vacant stares that normally await those forced to live with writers. We deeply appreciate their patience and wisdom.

Photo Credits

ONE

The Race

In the twilight at the Los Angeles Coliseum, sprinter Carl Lewis bounced on his toes, shook his arms loose, and then stepped into the blocks. From high in the stands, where Sam Ginn sat, you could pick out Lewis wearing a red USA singlet and shorts and sporting a distinctive square, brush-cut hairstyle. Slowly, the chants of "USA, USA" subsided and a tense quiet enveloped the stadium. It was a perfect southern California evening, warm and clear. A night for making history. Although the politics of the Cold War were ever present – the Soviet invasion of Afghanistan, the U.S. boycott of the Moscow Olympics in 1980, and now the absence of the Soviets and their allies from the games in 1984 – they seemed banished to the shadows tonight. Lewis hoped to vanquish those shadows by winning four gold medals, just as his idol Jesse Owens had done, triumphing over politics in Hitler's 1936 Olympics in Berlin. The 100-meter dash was Lewis's first test.

Rising with the crowd for the start, Sam Ginn watched the runners come to the set position. Tall and trim, and looking considerably less than his 47 years, Ginn relished the moment: the sudden quiet in the stadium, the flickering of the Olympic flame at the western end of the stadium. Like Lewis, Ginn had been born in Alabama. He was raised in Anniston, a small city that embodied many of the disappointed industrial dreams of the post–Civil War South. As a child in a working-class family, Ginn had dreamed of crossing over the tracks to the side of the city where big houses lined the boulevard and where people had the money to travel to events like the Olympics. In high school, sports gave him his first opportunity to enjoy a privileged status, and he had developed a taste for winning. Smart, intensely competitive, and personable, Ginn had reached Los Angeles by racing up the corporate ladder over the course of a 24-year career with the largest company in the world – American Telephone & Telegraph (AT&T).

Now a top executive with the newly launched enterprise known as Pacific Telesis Group (one of the seven "Baby Bells" created by the breakup of AT&T), Ginn was a guest at this event, hosted by the company he had spent all of his career serving. As a result of the breakup, AT&T had become Pacific Telesis's largest customer, competitor, and supplier in the marketplace for telecommunications; it was AT&T that had provided Ginn with tickets to this track and field event. Being wooed by his former employer, however, was only one of the many strange aspects of the world Ginn confronted in 1984. The settlement in the antitrust suit had completely restructured his economic setting, seriously disrupting his sense of mission and purpose in life.

For nearly a quarter of a century he and his peers in the Bell System had believed they were the public stewards of one of the greatest engineering feats in the history of mankind – the national telephone network. With the breakup, however, regulators and the courts had made it clear that competition, not Bell System planners, would shape the future. Ginn and his peers in the other Baby Bells were still trying to understand what kind of future lay ahead of them. Many were giddy with the prospect of entering new lines of business and developing the potential of new technologies. After all, they reasoned, their companies had access to enormous amounts of capital and tens of millions of existing customers. The capabilities of their organizations, as history had shown, were prodigious. Many people on Wall Street, however, feared that – because of their size and bureaucratic organizations – the Baby Bells would lumber down the track and finish far behind the fresh-faced newcomers to the telecommunications industry. Ginn favored neither of these two extremes. He was cautiously optimistic about Pacific Telesis and the industry. But in 1984 neither he nor the other Bell executives had a positive fix on their prospects, and none of them could possibly understand how much the past would shape their future.

That future was on Ginn's mind at the Olympics because he carried – while attending all of the events – a device that most people had never seen before. It was a brand-new $3,500 Motorola portable telephone. Known in the fledgling wireless industry as "the brick," the heavy, putty-colored device with a black face was one of the first portable cellular phones to be offered to consumers. It looked like it should be used by someone at NASA, not by a fan in the stands of the Los Angeles Coliseum. When Ginn turned it on, the phone transmitted radio signals to a tower that was part

Martin Cooper and the Motorola DynaTAC phone – While the Bell System focused on phones in vehicles, Martin Cooper and his team at Motorola concentrated on portability. The result was the DynaTAC, known in the industry as "the brick."

of the cellular wireless network Ginn's company had raced to complete in time for the Summer Olympics.

People stared and listened when Ginn made a call from his seat. They nudged their friends and pointed. Frequently, Ginn turned and offered them the phone. In a warm voice that still betrayed traces of his Alabama roots, he asked, "Would you like to call home?"[1] Ginn showed them how to punch in the numbers. Then inevitably they would shout to some distant friend or relative – inaugurating what would become one of the rituals of mobile telephony – the statement of location: "You won't believe this. I'm in the stands at the Olympics, talking on this radio phone." A Japanese businessman called Tokyo. Another man called his wife thousands of miles away. Each was delighted in a way that Ginn would never forget.

The mobile phone brought people together and gave them a new freedom. No longer forced to wait until they could "get to a phone" to share news with friends and family, people could reach out and touch someone from wherever they were. Here in the midst of the crowd in the Coliseum they could report on events live, and those listening on the other end could share in that exciting moment. Watching people talk on his phone and pass it through the stands, Sam Ginn had an epiphany. He recognized for the first time that these new phones represented more than an incremental improvement on existing technology. They created a whole new paradigm for communications. For years he had tried to imagine how Alexander Graham Bell must have felt at the beginning of the telephone era, at the realization that his new technology could become basic to the lives of millions of people. In Los Angeles, Ginn sensed that he was experiencing an "Alexander Graham Bell moment."

As the gun sounded down on the track, Ginn watched Lewis and the other sprinters bolt out of the blocks. Over the first fifty meters, Lewis lagged behind the two leaders, but then he changed gears and accelerated. Throwing his arms up at the finish, he won by nearly two meters. Ginn clapped with the crowd as Lewis took a huge American flag from a fan. He jogged through his victory lap beaming and waving to his cheering fans. Already the sprinter had embarked on the path to history.[2]

Ginn had a sense that day that he, too, was at the start of something big. But while a sprinter like Carl Lewis could look down the track, see his destination, and know what it would take to get there, Ginn faced a more uncertain future. It was obvious in 1984 to everyone in telecommunications that the industry was changing dramatically at the national and maybe even the global level. But it was not at all obvious what role wireless would play in this transformation. Nor was it clear whether Pacific Telesis or any of the other existing firms in the industry would be the leaders who would champion new technologies, styles of organization, or approaches to doing business. Over the next few years, it was sometimes hard for Ginn and his peers to tell what race they were running and who exactly was the competition.

For consumers around the world, however, wireless was an obvious winner, an innovation that introduced a whole new way of life. Today, there are more cell phones in use than personal computers, and increasingly people connect to the Internet through a wireless device rather than a PC. Awkward as it was, the Motorola brick was thus launching something extremely important for millions of people. It was not designed for astronauts. At the beginning of the wireless era in 1984, it allowed any ordinary person to do what had been a pipedream for decades – communicate with anyone, anytime, anywhere.

*

Consumer enthusiasm for wireless prompted a race to develop a new wireless world. The race, which is the central narrative of our story, took place amidst a series of dramatic transformations that occurred around the world in technology, politics, and trade, and we have tried to describe these transitions and the way they played out in wireless. The race was between entrepreneurs and their organizations, and we devote considerable attention to the nature of that entrepreneurship and its role in a long-established industry experiencing sudden change. Back in the darkest days of the Great Depression of the 1930s, when scholars of "the dismal science" were struggling to explain the collapse of the global economy as well as the factors that would lead the world back to happier days, Harvard University republished a study by Moravian-born economist Joseph Schumpeter. He was an unusual scholar who had many years before advanced a unique theory of economic development. Schumpeter's ideas about economic growth flew in the face of neoclassical and Marxist explanations. Innovation, he said, is the engine of economic growth, and the agent of innovation is the entrepreneur.[3] Capitalist economic progress was inherently uneven but entrepreneurs, if allowed to continue, would inevitably spur another surge of growth.

Schumpeter told his readers that entrepreneurs were not intellectuals, tinkers, or inventors. They were individuals in business who saw the potential for a new product, service, or process, or who perceived the opportunities in a new source of raw material, a style of organization, or a market. Entrepreneurs also had the temerity to act despite the serious challenges they faced. Uncertainty stemmed from a lack of reliable data. The path ahead was always unclear when innovation was taking place. Doing what was familiar was always easier than doing something new. Society normally resisted change, making it difficult for the entrepreneur to obtain capital. Banks were conservative. Established interests – public as well as private – that were threatened by innovation threw up roadblocks to change. Even if buyers could see the benefit of some new product or service, they often preferred what they already knew. To succeed, the entrepreneur had to overcome the resistance of individuals, organizations, and sometimes an entire society.

For the most part, economists, public officials, and the general public ignored Schumpeter's ideas for decades. He labored away, continuing to elaborate his theory and history of capitalism, but meanwhile economics was transformed in the late 1930s and 1940s by the ideas of John Maynard Keynes.[4] For several decades following the Second World War, the Keynesian model and national planning were the centerpieces of political economy in the capitalist democracies. In recent years, however, the Schumpeterian perspective on economic growth has experienced a remarkable revival as academics, business leaders, and policymakers – the neo-Schumpeterians – have sought to understand the dynamics of capitalism and the intricate relationships between economic growth and entrepreneurship in industries like wireless.[5]

Thanks in large part to the research of Alfred D. Chandler, Richard Nelson, and other scholars, neo-Schumpeterians recognize that much of the innovation that takes place in modern capitalism comes out of very large corporations. Schumpeter was suspicious of great corporate combines, not because they were monopolistic but because they tended to be organized along bureaucratic lines. Bureaucracy, Schumpeter said, was the enemy of the entrepreneur. It would stifle creativity. His ideal entrepreneur was the great individual who, in the nineteenth-century style, built and dominated an innovative business empire. It remained for a host of historians, economists, and business school analysts to show how, in the twentieth century, large corporations began to perform the entrepreneurial function. The most obvious manifestation of this major economic transition was industrial research and development (R&D), which yielded new products and new processes. Meanwhile, the top executives and managers of the corporations were organizing and reorganizing their enterprises in an ongoing effort to improve efficiency and foster further innovation. While this was happening, neither the old style of heroic entrepreneur nor the small innovative business disappeared; both remained important aspects of capitalist development. But they coexisted with large, multinational enterprises that achieved innovation through bureaucratic means.[6]

By political fiat, the modern wireless industry included both types of entrepreneurs. Early in the 1980s, the U.S. government made a fateful decision that would turn the development of wireless into a race between these two kinds of innovators. In every city across the country, two companies would be allowed to offer cellular telephone service. One would be run by people (like Sam Ginn) who had spent their entire careers working for AT&T or some smaller local telephone company. The other would be launched by individuals and organizations from outside the Bell System.

This experiment with a contrived form of dual competition would shape the future of the entire telecommunications industry and ensure that the grand legacy of the Bell System would not quickly wither away in the era after the breakup.

In the short history of the wireless industry, Craig McCaw was the kind of entrepreneur who delighted Schumpeter and still today makes American legends. An industry outsider from the Seattle area who risked everything to place the biggest bet possible on an emerging technology and an opportunity that he believed would change the world, McCaw and his team challenged the Baby Bells in the race for dominance in the U.S. wireless industry. McCaw became a major rival to Sam Ginn and the Baby Bells, and his vision and daring made him a billionaire when he sold out to AT&T. Confident that there would be a mass market because what customers wanted most was the freedom to roam, McCaw had a powerful influence on the fledgling industry.[7]

But Sam Ginn's story may tell us more about the development of wireless and the changes taking place today in global telecommunications. It also offers us insight into the future of other industries that were once state-run or highly regulated. Ginn and his managers at Pacific Telesis Group and later AirTouch Communications launched their revolution from the inside. Like their peers at the six other Baby Bells (Nynex, Bell Atlantic, Southwestern Bell, BellSouth, U.S. West, and Ameritech), they were handed cellular as a birthright with the 1984 breakup of the Bell System. Entrepreneurs like McCaw pushed them to become more flexible and competitive, and by responding successfully the former Bell System companies – Verizon, Cingular, and AT&T Wireless – emerged on top of the wireless world in the United States. At the end of 2001, these companies controlled more than 60 percent of the market.

Their main challenger today is an upstart British company called Vodafone. Led by Chris Gent, Vodafone has grabbed the inside lane in the wireless race by assembling an intimidating collection of assets on nearly every continent. With its acquisition of AirTouch in 1999, Vodafone began the race toward global consolidation while its main competitors were still focused on consolidating national markets. As a result of that acquisition, Gent had to integrate two very different corporate cultures and traditions. One was deeply rooted in the history of the Bell System; the other had succeeded by challenging the dominant telephone company in the British market. Fortunately for Gent, the two organizations shared a common strategy based on the assumption that wireless was the way of the future.

Sam Ginn – Sales and marketing were far more important to wireless than they had been to the Bell System. As CEO of AirTouch, Sam Ginn cultivated a new image that emphasized his role as the company's lead salesman.

AirTouch was a desirable acquisition because Sam Ginn had distanced himself and his organization from some, but not all, of the values of the bureaucrats and engineers who had long guided the Bell System. In order to become fast-moving, customer-focused, and innovative competitors, Ginn and his colleagues had abandoned crucial elements of the Bell way. They had made many mistakes, and they had never completely shucked off the

Bell culture. But their successful efforts to develop a new hybrid culture and organization had enabled them to create the largest wireless enterprise in the world by 1998. In the chapters that follow, we tell the story of this transformation – both personal and organizational – and explore the effects of this style of entrepreneurship on the wireless industry and the global economy.[8]

From this saga of late twentieth-century entrepreneurship, three other major themes emerge. The first involves technological change. Experiments with wireless telephones began nearly a hundred years ago, but it was not until the mid-1980s that this innovation was made available to large numbers of consumers. To a considerable degree, the wireless world we know today depends on the development of the transistor and the integrated circuit, innovations that have catalyzed the Third Industrial Revolution. These new technologies have driven change around the globe just as surely as waterpower and the steam engine propelled the British economy forward in the First Industrial Revolution of the eighteenth and early nineteenth centuries. Information Age technologies have fostered the development of the microcomputer and the Internet and built the great personal fortunes that successful entrepreneurship always yields.

Wireless is a distinctly Third Industrial Revolution industry. Today, miniaturization and increases in computing power are the basis for the new wireless world that fits in the palm of your hand. Although the impact of computers and broadband telecommunications has been discussed in hundreds of books, the social and economic transformations enabled by "anytime, anywhere" wireless communications are only just beginning to be understood.[9]

Another theme involves political change. Around the world, the failure of state-owned or highly regulated industries to sustain a high rate of innovation has led to deregulation and privatization. As a result, former monopolists in large industries, including telecommunications, have for some time been struggling to become effective global competitors. At the same time, regulators have been striving to redefine their roles as they shift from being watchdogs for consumers in a monopoly environment to managers of markets in a world of partially regulated competition. In California, where AirTouch originated, recent regulatory efforts to manage the transition to competitive utility markets have been highly criticized. But neither the critics nor the regulators seem to have noticed that most of the state's contemporary difficulties with electrical power were foreshadowed by the transition to competitive markets in telecommunications during the 1980s and 1990s. That is when business executives, politicians, and regulators all discovered, as Sam Ginn did, that it is difficult to leave the past behind. Despite enormous changes, the telecommunications industry and its political overseers are still deeply rooted in the engineering, managerial, and regulatory concepts of the previous century.[10]

The modern wireless industry became a global phenomenon very early in its development, and this provides our final major theme. Although corporations and nations have been trading in a global economy for hundreds of years, new transportation and information technologies and new policies toward trade have stitched together the global economy in a novel way, expanding the stage upon which political and economic institutions must act.[11] Sam Ginn and his team came onto this stage from a relatively naïve position as managers of a large-scale telecommunications company – Pacific Telesis Group – operating in only two American states, California and Nevada. In 1994, Pacific Telesis spun off its wireless operations to create AirTouch Communications, with Ginn as chief executive officer (CEO). By then, AirTouch had already become a global company with wireless networks and partners on four continents. But while AirTouch had become a global organization, it was not a traditional multinational corporation. It built its domestic and global enterprise in a new way, using joint venture partnerships and strategic technological relationships – firm structures characteristic of the most innovative Third Industrial Revolution industries.[12]

This book thus offers a personal, a corporate, and a general economic perspective on the wireless industry and a rapidly changing global economy. The study focuses on AirTouch and the efforts made by its managers to combine the technical and organizational virtuosity of the Bell System with the light-footed performance needed in wireless. This was not an easy combination to create. Around the world, former monopolists operating within national boundaries – including telecommunications giants like AT&T, Verizon, BellSouth, and SBC in the United States, as well as Deutsche Telekom, France Telecom, British Telecommunications, and Nippon Telephone and Telegraph – are engaged in a similar struggle to become global innovators.

Like telephone regulators and managers around the world today, the early wireless entrepreneurs were often baffled by events beyond their control and uncertain about the future of their enterprises. None of the wireless pioneers had a crystal ball. Many people believed in the early days of cellular that this new technology would succeed only as a convenience for the urban rich and powerful, a marginal business for existing wireline companies. The pessimists failed to see either the technological potential of wireless or the latent demand for a new form of communication. This was certainly the case at AT&T in the 1960s, when a generation of men and women were entering the Bell System looking for a ticket to the American dream.

TWO

Bell-heads

The Bell System had an enormous appetite for managers. Across the United States, hundreds of thousands of AT&T employees went to work every day and laid telephone cable underneath city streets, strung wire from poles along rural highways, maintained switches in central offices, answered calls for operators, installed telephone jacks in homes and businesses, delivered handsets and phone books to customers, assembled components for repeaters and handsets, calibrated microwave relays, and performed a thousand other tasks to keep the people of the United States in touch with one another and the world. Meanwhile, in the System's huge research laboratories, scientists helped pioneer many of the greatest technological breakthroughs of the twentieth century, including radio, television, and the transistor.

At all of the job sites related to these tasks, managers kept track of paper – enormous quantities of paper – as they documented the System's performance, including the activities of many employees in fifteen-minute increments. Bell System manuals set forth company policy on day-to-day operations and spelled out much of the daily routine required of the System's thick layers of management. Regulators who governed AT&T's virtual monopoly needed these records to justify the charges they allowed Bell System companies to levy on consumers who wanted to reach out and touch someone by phone.

AT&T, the holding company that controlled this massive organization, hired managers from all walks of life. Some were recent college graduates taking their first real job. Others entered management from the ranks of the System's blue-collar workers. In the early 1960s, almost all were men. AT&T's human resources program in those days constituted an elaborate and well-organized meritocracy that began with the company's recruiting system and extended to the grooming of the most senior executives.[1] During the decade that began with John F. Kennedy's campaign for president, the Bell System was an appropriate symbol for America's corporate prowess. It was a giant, vertically integrated, multidivisional organization run with an efficiency that was legendary when compared to other telephone systems around the world. Legions of middle managers, the "Bellheads" who made this System work, reported through elaborate chains of command to the president of AT&T at 195 Broadway in New York City.

Sam Ginn entered that system in the year that Kennedy was elected. Many of the men who would be his peers and competitors at the Baby Bells, as well as his colleagues at Pacific Telesis and AirTouch in the 1990s, began working for AT&T during that same era. Over the next 24 years Ginn, like the rest of his cadre, learned how to thrive in the highly regulated, highly structured world of the Bell System. At that point in its history, it was a system that had not known competition for nearly half a century. But that was about to change.

Compared to the rest of his cadre, Ginn didn't have a background that made him stand out as an obvious winner. As a boy he had thrived on competition and public performances, but he was not a remarkable student. To satisfy his parents, Harold and Myra Ginn, he maintained a B average, but he put most of his considerable energy into basketball, baseball, cheerleading, and theater. His peers voted him the most popular boy in the class, but the principal of Anniston High, Mr. Nash, was not impressed. When it came time for Ginn to apply to college, Nash refused to write him a letter of recommendation. He said, "I know how hard your dad works, and I know how you spend your time. You're going to go away to school, spend a lot of your dad's money, flunk out, and be back here in nine months." Ginn swore he would not disappoint the principal or his parents, and eventually he got the letter he wanted.

During his first year at Auburn University, however, Ginn nearly fulfilled the principal's prophecy. Ill prepared for college-level academics, he struggled to keep his head above water. In the fall, he got Cs and one B, but in the winter quarter his grades collapsed. He flunked trigonometry, got a D in English, and barely held on for Cs in his other classes. Anniston High had been gentle and supportive to its popular sports hero. At Auburn in the late 1950s, he was laboring within a draconian academic regime in which professors routinely told students on the first day of class: "Look left and look right. One of you will fail or drop out by the end of the quarter."

Sam Ginn struggled to turn his academic career around. For the first time, he applied to his schoolwork the discipline he had previously reserved for basketball. He engaged the competitive environment his professors created, and that effort paid off. By the time he graduated in 1959, he had become a member of the Blue Key honor society. Academic success fueled his confidence, but it didn't help him decide what he wanted to do with his life.

His impatience and the need to help pay for the cost of his college education accidentally brought him into the field of telecommunications. During his first week at Auburn he had joined the Reserve Officers Training Corps (ROTC) to earn an extra $40 a month for school. The counselor sent him to the armory to register, but when Ginn got there the building was teeming with young men. He picked the shortest line, one where an image of two crossed flags, one red and one white, marked the station. Moments later, he had enrolled in the Signal Corps.

As a result, he spent several weeks every summer during his college career training in the blistering heat at Fort Gordon, Georgia. He set up field radio and PBX systems during military exercises and got his first experience with wireless communications. After graduating from Auburn, Ginn entered the Army as an officer in the Signal Corps. After more training at Fort Monmouth, New Jersey, he was assigned to the Third Army's Headquarters in Atlanta – the Army's global communications center. If he had followed the script for the American myth of success, he would have decided right then that his future would be in telecommunications. But he was still undecided on a career when he got out of the Army, still not sure what he wanted to do with his life. He interviewed with a dozen companies. Not until he talked to the father of a friend from ROTC did he decide to go into telecommunications, starting as a Bell management trainee.[2]

Over the years, the Bell System had created one of the most sophisticated management recruiting, screening, and development programs in American industry.[3] When Sam Ginn was hired, the company was making further changes to increase retention and accelerate the promotion of high-quality management trainees.[4] The company screened for candidates who ranked in the top 25 percent academically and had some leadership experience. Under this new regime, recruiters administered a battery of SAT-like intelligence tests to potential candidates. But the crucial part of the screening was the interview. Under the influence of psychologist Frederick Herzberg, the company was trying to separate "work motivated" team players, who wanted recognition, advancement, and achievement, from "maintenance" individuals, who had a tendency to focus only on immediate personal gratification from their environment.[5] This system played to Ginn's strengths as a competitive and internally motivated team player. He scored high, and AT&T offered him a job. Out of thousands who had applied, Ginn had been recruited into an elite group of thirteen headed for the fast track.

Ginn's experience in the AT&T management system began with an elaborate three-month indoctrination program at the regional headquarters of Long Lines, the Bell System's long-distance division. Sporting crew cuts and business suits, white shirts, narrow ties, and argyle socks, Ginn and his peers arrived in Cincinnati. Having graduated from college, he was in the majority of Bell System recruits. Women had not yet entered the management ranks of the Bell System in great numbers, though they dominated some segments of the business, including operator services and the secretarial staff.[6] Like about half of his contemporaries, Ginn was not married. Like the majority of his peers he was a registered Republican, but he came from a background that had more in common with that of the blue-collar men who had earned a shot at management by their performance on the job. His father had not graduated from high school and had never held a white-collar job. Ginn was new to the middle class and to management.[7]

The training focused on Bell System practices. For every managerial action there was a Bell System way to perform. Ginn learned how to fill out expense vouchers, vacation schedules, time sheets (in fifteen-minute increments), safety reports, and dozens of other forms. He role-played with other trainees, taking turns being the boss and the worker. It was an elaborate process, the cost of which was of course folded into the rate base that regulators used to figure AT&T's tariffs. Compared to rural Alabama, the well-lit, clean, and modern offices of AT&T were amazing, and Ginn was impressed with the discipline involved in every aspect of the company's operations.

After Cincinnati, Ginn returned to Atlanta and rotated through a series of jobs designed to familiarize him with the firm and the work of running a telephone company. Under this regime, each trainee was quickly put into a supervisory role to test his management skills. Each was assigned a mentor who provided a sounding board and monitored his performance. From the first weeks in Cincinnati, the trainees were constantly told how special they were and how much was expected of them. Everyone they encountered in the company knew they were on a fast track. But they knew, and everyone else knew, that they had only five years to prove that they were middle management material. Otherwise they would be back out on the street, looking for a new job.

Ginn thrived. Without a waver or serious slip, he stayed on Bell's fast track. That meant frequent moves from one kind of job to another as the company pushed him to see what his limits were, gave him a broad range of managerial experiences, and groomed him for the next step up. He climbed telephone poles to replace rotten crossarms in Georgia. In New Orleans he had his first experience with sales and liked it. He also enjoyed the internal competition, and he liked to be creative in solving customers' problems while making money for the company. His supervisors at AT&T were pleased. One of them, Charlie Brown (Ginn's division boss in Atlanta), became an important mentor – and a mentor was a good thing to have in the Bell System, especially one who was on the fast track himself.

A network engineer by training, Brown had virtually grown up in the Bell System. Both his parents were mid-level telephone company employees. He had dug ditches for cable during the summer when he was 18. After college and service with the U.S. Navy in World War II, he went to work for AT&T. Reflective and soft-spoken, he had a composure that won him the respect of everyone who worked with him. As his intense blue eyes suggested, there was a great deal of energy bottled up in his quiet manner. He too was a natural competitor, a high achiever, with a special knack for engineering and operations. Ginn and Brown got along well and understood one another completely. Over the course of the next two decades, their paths would cross a number of times.

Ginn married while he was mastering the System, and he, his wife Ann, and their growing family moved nearly every eighteen months over the next nine years. That was merely one of the tests Bell managers and their families had to pass to keep moving up. Fortunately for Ginn, he had fallen in love with a woman who understood the System. Ann Vance had grown up in Winston-Salem, North Carolina, where her father had spent most of his career as a mid-level manager with Southern Bell Telephone. She and Sam agreed that he should reap the rewards the System bestowed for staying the course. One was a front-row seat on the conflict brewing between the Bell System and the government. This was a struggle that would eventually force Brown, as CEO of AT&T in 1982, to make the historic decision to break up the System.[8]

✻

From the point of view of Sam Ginn, Charlie Brown, and most AT&T executives, the strength of the Bell System lay in its centralized engineering and management. Competition among parallel networks in the telephone business made no sense to the Bell-heads. They believed – with a sincerity that was astounding to those outside the organization – that the System existed to serve the public interest, and they subscribed to the company's two overriding goals: universal service and superior quality. Earnings were important, but only to those executives who held the highest positions in the Bell System. In the day-to-day world of managers, the service goals trumped everything else. They had been articulated decades earlier as part of an agreement with the government of the United States that in effect recognized what many believed was "the natural monopoly" of telecommunications.[9]

Following the invention of the telephone in 1876, the Bell System had gradually evolved a vertical and horizontal structure that integrated most of the basic activities of telecommunications in the United States within a single company.[10] Western Electric, an AT&T subsidiary, manufactured the telephones and equipment needed to keep the network running.[11] The Bell Operating Companies provided local telephone service in cities and communities across the country. The Long Lines Division ensured that customers in distant cities would be able to connect with one another.[12] Meanwhile, the world-famous Bell Laboratories focused on the scientific and technological research that sustained innovation within the System.[13] Overseeing all of these operations were the AT&T General Departments at the company's headquarters in New York. Most of this historic integration had taken place in the first half-century of the Bell System's development, and president Theodore Vail had shaped the basic strategy and culture of the System following its reorganization and revitalization in 1907.[14]

Competition had been tried in the telecommunications industry in the United States. After Bell's initial patents expired in the mid-1890s, entrepreneurs started rival telephone companies to compete with the System's local operating companies.[15] Intense price competition left the System weighted down with debt and on the brink of bankruptcy by 1907. Rescued and reorganized by investment banker J. P. Morgan, the System espoused a new strategy, the Vail strategy, that included the explicit goal of universal service. With its local and long-distance operations consolidated under the single corporate structure provided by AT&T, the organization vigorously mounted a strategy of acquisition to expand its control of the market for telephone services.[16] After the U.S. Justice Department challenged AT&T's expansion as a violation of the federal antitrust law, the company and the government reached a historic agreement in 1913.

Known as "the Kingsbury Commitment," this agreement allowed AT&T to gradually consolidate a virtual monopoly over local telephone service in exchange for a promise to connect independent companies to the public switched network, accept government regulation, and pursue the goal of universal service.[17] Essentially, the United States government, like governments in Europe and Asia, had decided that competition in telecommunications was not good for consumers because it led to duplicate networks and inefficiencies that diminished the potential value of telephone services to customers.[18]

To maintain service quality, the government also accepted AT&T's argument that others should be precluded from attaching their own equipment to the network. This decision guaranteed AT&T a dominant position in telephone equipment manufacturing in the United States, a position that periodically made state regulators and government officials in the Justice Department uncomfortable – especially the regulators in California. Government lawyers and regulators worried about the potential this gave AT&T to use its manufacturing arm, Western Electric, to pad its revenues.

In 1949 the government launched a suit aimed at forcing AT&T to divest Western Electric. That suit had been settled in 1956 with a consent decree that allowed AT&T to keep its vertical structure in exchange for a promise to stay out of noncommunications-related businesses and to license its patents freely to anyone who wanted to make use of them. AT&T thus had to forego a major opportunity to capitalize on its invention of the transistor and, incidentally, to profit from manufacturing radio communications equipment such as mobile phones. AT&T's leaders believed that this settlement would protect the firm's core business – telecommunications – from competition for many decades to come. They were wrong.[19]

During the 1960s, when Sam Ginn was just getting started in the System, the relationships between the government and the phone company were once again growing uneasy. Regulators at the Federal Communications Commission (FCC), lawyers in the Justice Department, and judges in federal courts became increasingly ambivalent about AT&T's monopoly, and would-be competitors emerged in three important sectors: long-distance transmission, telecommunications equipment, and mobile telephony.

Technological pressures against the monopoly had been mounting for some years. As early as the 1940s, entrepreneurs in the mobile field had begun to urge the Bell System and the government to accept competition. The FCC provided a separate set of frequencies for these businesses, known as radio common carriers (RCCs).[20] Then, in the 1950s, the development of microwave provided the first alternative to cables for long-distance transmission. Startled by the success Philco and others had with microwave, AT&T launched a crash program to develop its own systems. At the same time, AT&T pursued a political strategy aimed at keeping other microwave innovators out of the transmission market. Bell convinced the FCC to grant microwave transmission licenses only to common carriers. In light of this decision, Philco and others at the cutting edge of the technology withdrew from the industry. Although some private-line development continued as right-of-way companies like railroads and pipelines installed systems for their own use, and even though some private transmission systems were allowed to provide television in rural areas, the pace of technical change was clearly retarded. The Bell System gained time to catch up.[21]

AT&T's regulatory tactics could not, however, keep others from the microwave business for long. The opportunities to reap entrepreneurial profits kept attracting would-be competitors. Motorola, the leading manufacturer of microwave equipment in the United States in the 1950s, urged potential customers to pressure the FCC to allow private-line microwave communication in competition with the Bell network. The FCC's "Above 890" decision, issued in 1959, approved this new use.[22] Suddenly, the market for transmission equipment opened up.

Each crack in the monopoly encouraged entrepreneurs to press harder to break down the political barriers to competition. The next pressure point was long-distance microwave transmission. In 1963, Microwave Communications, Inc. (MCI), a fledgling organization with very little capital but a great deal of political and legal savvy, applied to the FCC for permission to construct eleven microwave towers to provide private-line service to corporate customers between Chicago and St. Louis. MCI promised to make it easier for customers to hook up their modems and equipment and to deliver dedicated service at prices that were lower than those of AT&T. To Sam Ginn and his Bell colleagues, it was obvious that MCI was proposing to "skim off the cream," taking the profits from a lucrative long-distance route without paying for the hardwired, local loops that cost much more to build and maintain. MCI claimed that it could cut prices because its equipment costs were less. But no one in the Bell System believed that argument, and AT&T (along with Western Union) opposed MCI's application. Despite this opposition, the FCC finally granted MCI its first construction permits in 1969. *Communications* magazine told readers that the decision "stands to revolutionize the communications structure of the country."[23]

About the same time, another crack in the Bell monopoly developed as a result of a case related to mobile phones. In Texas, Thomas Carter had invented an acoustical device that allowed a mobile radio customer to be patched into the Bell System network. AT&T refused to allow customers to use the "Carterfone," asserting Bell's responsibility for "end-to-end" connection over its network. When AT&T discontinued service to

Carterfone customers, Carter filed an antitrust suit. The court referred the issue to the FCC, and in 1968, in a significant shift from previous decisions, the FCC overruled AT&T's objections. In doing so, the Commission effectively opened up the market for customer premises equipment as long as AT&T could not establish that it was harmful to the network.[24] That proved difficult to do, and the FCC was now edging toward a new position, willing at last to consider introducing competition at various margins of the business. Over the next fifteen years the margins kept moving, and Bell loyalists like Sam Ginn and Charlie Brown found themselves caught between the System they were still trying to run and a flood of competitors rushing to take advantage of the monopoly's crumbling walls.[25]

By 1972, AT&T could anticipate increasing competition in both equipment and long distance. MCI had its service up and running, and that year the company went to the stock market to raise more than $100 million to finance a communications network that would encompass 165 American cities.[26] Over the next several years MCI argued with AT&T, regulators, and the courts over whether it had a right to connect its private lines directly into AT&T's public switched network. AT&T said no. With interconnection, AT&T said, MCI would no longer be supplying private-line service but rather standard, message telephone service. MCI would be able to provide this service, AT&T charged, without having to pay the subsidies to local telephone systems that the FCC required of AT&T. MCI countered that its service would be provided only to a small number of exclusive customers and therefore would pose no threat to the giant Bell System. But private-line service was less profitable than message telephone service, and MCI fully understood that fact. Within a matter of years, this struggle led to the antitrust suits that would bring down the Bell System and have enormous implications for the telecommunications industry around the world.[27]

When 36-year-old Sam Ginn arrived in New York in 1974 as AT&T's youngest vice-president, he faced the MCI challenge head-on. Placed in charge of rates and tariffs for private-line services, he soon discovered that MCI and its entrepreneurial CEO, Bill McGowan, had big ambitions. MCI filed a proposal with the FCC for a new service called "Execunet." Under this proposal, MCI would essentially offer regular, metered long-distance service to all comers. Recognizing this new threat to the System, Sam alerted his superiors. They sent him to Washington, DC, where he spent a great deal of time testifying over the next year as the FCC evaluated MCI's proposal and as AT&T repriced its own private-line services to meet this competitive threat.[28]

The negotiations and hearings over AT&T's new pricing structure forced Ginn to become an expert in the sometimes bizarre and always frustrating world of regulatory politics. With AT&T sitting on one side of a table, the FCC staff on the other, and interveners and competitors scattered around the room, Ginn ground away in an arcane debate over AT&T's costs and prices. Ultimately, a new pricing structure emerged. But as Ginn realized, AT&T's prices were socially based, driven by policymakers' historic efforts to extend low-cost telephone service to the vast majority of Americans. In the transition to competitive pricing, the socially based price system would some day be lost.[29]

Convinced that the Bell network was a great national asset and that more would be lost than gained through competition, AT&T's leaders dug in their heels. From Sam Ginn's perspective, they were doing the right thing. But AT&T's refusal to allow MCI to connect to the public switched network left the System under fire from both private and public antitrust suits. The Bell System was at risk, but nothing happened immediately. Over the next four years, both cases moved slowly through the process of document production and pretrial motions. When Judge Harold Greene was appointed to hear the federal case, however, he made it clear to both AT&T and the government that he intended to move more quickly to trial. Pressure on the Bell System increased when MCI won its case in June 1980 and received a $600 million judgment against AT&T (which tripled to $1.8 billion under the terms of the Sherman Antitrust Act). AT&T began to consider settlement options more seriously. Over the next year and a half, AT&T and the Department of Justice explored a number of scenarios.

Paradoxically, the pressure on the Bell System increased with the election of Ronald Reagan as president in 1980. After the Reagan Administration appointed Stanford University law professor William Baxter as the new assistant attorney general for antitrust, the need to settle out of court became even more pressing for AT&T. Baxter looked upon the Bell System as a government-protected monopoly, and he promised to litigate the AT&T case "to the eyeballs." Baxter was not particularly concerned about Western Electric and Bell Labs, the vertical components of the System. But he was certain that the System's horizontal integration gave AT&T too much market power. His "elegant solution" called for AT&T to divest the local Bell Operating Companies, which would continue to be thoroughly regulated by both the FCC and state regulators as bottleneck monopolies.[30]

AT&T initially rejected the elegant solution. But by the end of 1981, Charlie Brown believed that it was the only way that AT&T could settle with Baxter, break free from the constraints of the 1956 Consent Decree, and develop the enormous technological capabilities of the Bell System. The "elegant solution" would also provide the best possible deal for the stockholders. Nevertheless, few outside of the inner circle of negotiators anticipated this decision. When Brown and Baxter announced in January 1982 that AT&T and the Department of Justice had agreed to break up the Bell System, they stunned the world. They also changed the world in decisive ways. The new competitive order that emerged set the stage for the birth of wireless and for a global transformation of telecommunications.

THREE

Birth of a Wireless World

Phil Quigley was not a typical Bell-head. Instead of studying engineering, he took a degree in marketing. When he graduated from college in 1967, he followed a classmate's suggestion and applied for a job at the phone company even though he had serious doubts about working in the Bell System. To Quigley and anyone else in marketing, the phone company was "a nonentity." Phone service was something you took for granted, like water coming out of the tap or power from an electrical outlet. It was something that was there when you needed it. You didn't think about where it came from, and the Bell System had not devoted much energy to marketing its service. But this friend said the phone company had a management training program, so Quigley decided he would take a flyer.

When he interviewed with Pacific Telephone and Telegraph, he learned that the company was trying to invigorate its Yellow Pages operation and was looking for college graduates to do the job. He accepted an offer, went to work, and quickly discovered that Yellow Pages functioned as a world apart in the Bell System. For one thing, it faced substantial competition from radio, television, newspapers, and other advertising mediums. Sales determined success, yet when Quigley talked to people who were in the mainstream telephone operations, "they were reluctant to set sales quotas." As a franchise monopoly, the local phone company was sensitive to charges by consumer groups and especially by regulators that they were overselling customers.[1]

Quigley thrived in the Yellow Pages division and, unlike most Bell System executives, did not leave – at least not until 1982. He did not rotate through the various divisions of Pacific Telephone and Telegraph, nor did he move around the country as Sam Ginn did in the Long Lines Division of

AT&T. In 1972 Quigley moved to New York for a three-year tour of duty at AT&T's headquarters, but he stayed in Yellow Pages. In New York, he could for the first time see that business from a national perspective. He also saw the centralized planning of the Bell System in action as he worked on a team to redesign Yellow Pages operations around the country. As with most of the System's managers, his tour of duty at headquarters signaled that Quigley was a man on the rise. Within five years of returning to California and Pacific Telephone and Telegraph, he had been promoted to head a Yellow Pages operation producing more than half a billion dollars in revenue annually and growing 20 percent a year.

Then everything changed. In Washington, Charlie Brown and William Baxter announced the plan for the breakup. Months later, Quigley learned he had a new boss. He would now be the number-two executive in California's Yellow Pages. No one took the time to explain what he had done to deserve his demotion. Quigley sensed that a bigger change was coming, but he had no idea what it was until September of 1982 when Gil Sheffield, one of the officers of Pacific Telephone, called him in for a meeting.

Sheffield explained that, as part of the division of Ma Bell's assets, the Baby Bells were to inherit a new, yet-to-be-launched business called cellular communications. The company did not see this new enterprise as a technology-oriented business. Instead, Sheffield said, it would be driven by marketing and sales. As one of the few people in Pacific Telephone who had actually run a competitive business and been responsible for the bottom line, Quigley was the obvious candidate to lead this cellular initiative. Sheffield asked if he would be interested.

Quigley laughed. "I don't know a thing about it," he answered, reminding Sheffield that he was neither a telephone man nor a network engineer. "But it sounds good." Soon Quigley was headed east again, back to AT&T's headquarters, to begin learning about wireless communications – a technology that had been incubating for nearly ninety years.

*

The incubation had begun in Italy, where Guglielmo Marconi, a young Italian inventor, was the first person to believe in the miracle of wireless transmission. The son of a wealthy Italian father and an Irish mother, Marconi read in an 1894 electrical journal about the experiments of Heinrich Hertz, the first person to produce and detect electromagnetic waves. Fascinated with the idea of using these waves for wireless telegraphy, Marconi began experimenting in two large rooms at the top of his parents' villa. Refining Hertz's apparatus, Marconi figured out how to receive Morse code messages over a distance of two miles, and in 1896 he took out a patent on his inventions.[2]

Determined to be an entrepreneur as well as an inventor, Marconi soon received his first lessons in the resistance that every innovator encounters. He was unable to interest the Italian government in his experiments. Undeterred, Marconi took his ideas to England, where his mother's connections to the Irish aristocracy helped the 22-year-old would-be entrepreneur gain introductions to government officials and some capitalists who were intrigued by the potential of wireless telegraphy. Marconi demonstrated a device that would send messages up to eight miles. His success led to the formation of the British Marconi Company in 1897, and the British Admiralty installed Marconi's equipment on some of its ships.

Marconi believed that wireless transmission would quickly become a low-cost competitor to fixed-wire and transoceanic cable transmission of telegraph signals. Focusing on increasing the range of transmission, he was able to send a signal 31 miles across the English Channel by 1899. At that time many scientists believed that the curvature of the earth would limit the range of wireless transmissions to 200 miles or less. But Marconi soon proved them wrong. Flying a kite for an antenna on the coast of Newfoundland on December 6, 1901, he was able to hear a faint signal of the telegraphic letter "S" transmitted from his sending station in England 1,700 miles across the Atlantic.[3]

Despite his improvements in the technology, Marconi's efforts to profit from his inventions hit a wall. Both public and private interests mounted opposition to his innovation, a response that Phil Quigley and Sam Ginn would encounter repeatedly in California. Frequently the private interests (with investments in fixed-line telegraphy) resisting Marconi's innovation were able to work through the government; this, too, was part of an almost universal pattern of reactions to the threat of entrepreneurial competition. Admittedly, wireless at that time was still vulnerable to interruption by atmospheric conditions and was thus an inferior technology where cables existed. But it was a new technology with considerable promise for further improvement, hence the fear of change and competition.

Marconi aggressively fought on, but he proved to be a better inventor than entrepreneur. He tried to preserve a monopoly over the new technology, but in doing so he alienated important potential customers. Meanwhile, German and American competitors emerged and improved on his components. As a research director, Marconi suffered from myopia. Focusing tightly on developing the technology for wireless telegraphy, he failed to see the potential for the development of a radio telephone. For fifteen years, he and his backers struggled, unable to turn a profit. They were not alone. Over the next half century, many would experiment with wireless transmission but fail to develop effective radio telephony. One result of their efforts would be a thriving radio industry that would ensure Marconi's place in history. But telegraphy would lose out not to the radio but to the hardwired telephones that increasingly dominated the world of point-to-point communications.

This would be the case even though other inventors continued to make technological progress in wireless. While Marconi struggled, his American rivals – Lee de Forest and Reginald Fessenden – soon passed him and took the lead in the wireless field. A professor at the University of Pittsburgh, Fessenden had abandoned his position in 1900 to experiment with wireless transmission of speech. That same year, he succeeded in broadcasting a voice between two antennae a mile apart using spark transmission technology. Recognizing the limits of that device, he turned to Nikola Tesla's research and began experimenting with the idea of continuous wave transmission. Fessenden was convinced that he could design a system capable of carrying wireless telephone conversations across the Atlantic Ocean. What he invented, instead, was the broadcast radio industry. On Christmas Eve in 1906, he transmitted a program of music and speech that was picked up by amazed shipboard wireless operators in the North Atlantic.[4]

While he was improving his broadcast techniques, Fessenden continued to experiment with wireless telephony. In July 1907, he successfully transmitted speech from Brant Rock, Massachusetts, to Jamaica, Long Island, a distance of 180 miles. But few people were impressed with his accomplishment. J. J. Carty, the Bell System's famed chief engineer, tried Fessenden's system and reported that the voice was very faint, too faint for commercial development. The transmission was frequently interrupted by "atmospherics" or static. Given the opportunity to buy Fessenden's patents and the company formed to develop them, the Bell System declined. A practical outcome from long-distance wireless voice transmission seemed too far in the future to justify even a modest investment.[5]

Whereas AT&T passed on the idea of radio telephony, stock promoters were less risk-averse than the Bell System's engineers. By the end of the first decade of the twentieth century, wireless stocks had proliferated as companies promised to let investors in on the ground floor of an industry that they claimed would soon be as important as the telephone system. It was the e-commerce of the day. One startup launched in 1907 aimed to promote the inventions of a young American named Lee de Forest.[6] De Forest had written his Ph.D. thesis at Yale on Marconi's obsession, the potential for wireless telegraphy. After graduating in 1899, he continued to work on the technical problems of wireless transmission. In 1906 he developed one of the most important inventions in the history of electronics, an instrument he called the "audion": a three-element vacuum tube that could send and amplify a continuous radio wave.

Like Marconi, de Forest was more talented as an inventor than as an entrepreneur. He had trouble perfecting his audion, and after the government began investigating wireless stock promoters in 1910, de Forest's Radio Telephone Company slid into bankruptcy. Eventually, he was forced to sell the rights to the audion (as a repeater for telephone lines and for radio) to AT&T. He then abandoned the field of radio telephony.[7] Engineers in AT&T's fledgling research branch improved on de Forest's invention and created the "high-vacuum tube" for amplifying sound in a telephone cable. Using this new device, the repeater, the Bell System was able to span the country with long-distance lines and open transcontinental service in 1915.

The high-vacuum tube also opened the path to wireless telephony for the first time. In their 1915 experiments, AT&T's engineers were able to send a telephone signal through the airwaves from Darien, Connecticut, to the Panama Canal Zone, 2,100 miles away. Atmospheric disturbances made the sound quality poor, and in subsequent calls to Europe only a few faint words could be heard. AT&T concluded that its enormous investment in a hardwired network was safe and that the future of wireless telephony was limited to those few inaccessible places where wire could not be strung. One of those places was Santa Catalina Island, off the coast of Los Angeles, where Pacific Telephone inaugurated the first commercial radio telephone service in the world in 1920.[8]

Although the Bell System's management decided that radio telephony had limited utility, others were encouraged by the opportunities they saw to use shortwave, local transmission for the military and police. Two Detroit policemen developed the first automobile radio receiver in 1928. It allowed dispatchers to transmit one-way radio calls to police cars, sending them speeding to one of the many crime scenes in the Prohibition decade after World War I.[9] Bayonne, New Jersey, deployed a two-way system for the first time in 1933.[10] Shortly before the United States entered World War II, the New York Telephone Company began offering the first commercial service to public utility companies for emergency dispatching in Manhattan and Brooklyn.[11] None of these systems met the everyday needs of ordinary citizens, and for most people in this era the ordinary black landline telephone still represented the modern age.

During World War II, government researchers put radio technology to work in the cloak-and-dagger world of espionage. Agents working for the Office of Strategic Services (the predecessor to the CIA) in occupied countries carried portable and easily concealed "Joan-Eleanor" radios that they used to direct the bombing missions of Allied aircraft. Motorola developed a two-way radio that used FM rather than AM frequencies, a unit that quickly became known as the "walkie-talkie."[12] Toward the end of the war, the FCC licensed experimental mobile dispatching systems for taxi companies, and this stirred up further demand from truckers, construction contractors, and delivery companies.[13] By war's end the public was fascinated with this new technology, and in 1946 cartoonist Chester Gould gave his hero, Dick Tracy, a two-way radio wristwatch that enabled the ace crime fighter to blast away at public enemies while calling for help.[14]

As demand for mobile telephone service increased, the Bell System test-marketed a public system in the St. Louis metropolitan area.[15] But there were glitches. The equipment was not sophisticated enough to prevent interference from adjacent channels, so Bell had to cut back to just three frequencies and could thus serve only a few customers. A year later, the company tested a similar system along the highway between New York and Boston, but there was too much interference and too many holes in the coverage. Grinding away at the technology in typical Bell fashion, Western Electric and Bell Labs managed to iron out these problems and enable the System gradually to extend service to other markets.[16]

Thinking that the demand for mobile service could go higher, the company had asked the FCC in 1947 to allocate a large number of frequencies for mobile service. But the FCC reacted conservatively. The Commission rejected the Bell proposal on the grounds that mobile phone service was a luxury. The agency preferred to preserve these frequencies for more important uses, including those of the police, fire departments, and other "public interests."[17]

Meanwhile, the FCC had authorized frequencies for use by a rag-tag group of companies that were competing with the mighty Bell System in providing this service. Launched in many cases by men who had served in the Signal Corps during the war, these radio common carriers (RCCs) provided private service to customers.[18] Through the 1950s, the mobile market was divided into three parts: taxi-style private dispatch services; private networks; and the Bell System's public mobile network. The Bell network was by far the smallest of the three.

During these early years, the RCCs introduced a number of innovations. Bell System customers, for example, had to be connected to the public switched network by operators. However, an Indiana entrepreneur named Ramsey McDonald, working with a local independent phone company, was able to develop automatic equipment that allowed customers to dial straight into the public network. AT&T representatives went to look at McDonald's operation in October 1957 but did not introduce their own automatic dial system (known as Improved Mobile Telephone Service) until 1961.[19]

Customers who subscribed to the mobile service in this early era put up with a lot. The early car phones came in boxes that were so big they had to be installed in the trunk of the car. The size of the phone units shrank somewhat as the technology improved, leaving room in the trunk for a suitcase or two. But in the 1950s, most customers still connected to an operator who placed their call, and the systems still operated in half-duplex mode – meaning you could not listen and talk at the same time. You had to press a button to talk and release it to listen.

Improvements in the technology were probably retarded by the 1956 Consent Decree that the Bell System entered into with the government. Under the terms of this agreement, AT&T had to abandon the manufacturing of mobile radio equipment.[20] While it continued research in the field, the Bell System now had to develop the technology primarily by working with Motorola, which became the System's largest supplier of mobile network hardware and subscriber equipment.[21]

Motorola, under contract to AT&T, pushed ahead with improvements. One was full-duplex service, which allowed customers to talk without having to press a button to switch back and forth between listening and speaking. This became standard in the mid-1960s using Motorola's fully automated system. Crystals and tubes later gave way to transistors and then microprocessors and integrated circuits, improving the quality of the transmission.

Despite these improvements, however, little could be done to expand the number of channels available to customers.[22] Mobile radio service still relied on a single high-powered station to serve an entire city. A given channel could only handle one call at a time. The only apparent way to meet increases in demand was to add channels from the limited radio spectrum. But that sent advocates of mobile service into combat against broadcasters, Defense Department officials, and other users who fought any effort to deprive them of spectrum. In 1968 the FCC proposed allocating additional frequencies for mobile and portable radio systems, but the agency's request for comments and proposals on how to use these frequencies quickly deteriorated into a long and rancorous public debate.[23] Even the largest private corporation in the world could not drive the process of innovation through this hedgerow of organized opposition.

As a result, by the 1970s the demand for mobile service had far outstripped the capacity of the system. To Sam Ginn and other managers mastering the Bell hierarchy, wireless thus remained just a curiosity, a novel technology that had little to do with the national network that was the System's greatest accomplishment and strongest claim to public support. Nationwide, there were only 140,000 customers, and for each one there was another on the waiting list. In the New York metropolitan area, with 14 million registered vehicles, the system could accommodate fewer than 600 users – and only twelve of them at any given moment. People who wanted mobile phone service had to wait for an existing customer to retire or die. When they did get a phone, they had to place the call and then wait for a green light to signal that an open channel was available. Often, they never got the green light. It wasn't hard to understand why many customers who had service rarely made a call.

A hint at the underlying demand for mobile communications came in the 1970s when there was a sudden national craze for Citizens' Band radios. At first, CBs were popular primarily with truck drivers who used the radios to break up the monotony of a long haul and to keep one another informed about fuel locations during the Oil Crisis and law enforcement along the highways. But in the late 1970s, CBs became wildly popular among the general public, and the FCC was nearly overwhelmed by as many as a million applications a month for licenses.[24] The CB craze finally alerted a number of people in the industry and at the FCC to the desire for a popular form of wireless communication. As one industry observer put it, the CB excitement showed "a powerful impulse on the part of every element in our society to communicate ... even sometimes when they have little to say."[25]

Other hints appeared in the growing market for paging services. The paging concept had originated with Charley Neergaard, a medical supply salesman who was hospitalized for two months in 1939 and driven to distraction by constant calls for physicians and other personnel over the hospital's loudspeakers. Neergaard pitched the idea to Al Gross, a radio pioneer who had developed a wireless system for the military during World War II. Gross built a pager prototype in 1949 and sold the rights to Motorola in 1952.[26] Over the course of the next decade, Motorola worked on perfecting the technology and defining the market.

Paging was fundamentally different from traditional landline or mobile radio telephone services because it defined the individual as the destination of the communication, rather than a telephone handset located in either a building or a car. Thus it set the stage for the future development of the Motorola "brick" and other hand-held portables. Yet Motorola initially conceived of paging as only a private network operating within a single building or complex like a hospital or factory. It was not designed initially to operate on a citywide basis.

That concept changed, however, in the 1960s, when Motorola introduced the first VHF pagers designed for use on networks that could accommodate several thousand users over a wide area. Instead of promoting this system solely for private use, Motorola began to encourage the RCCs to establish subscription paging services.[27] Motorola's efforts fueled the growth of the paging business and also expansion of the RCCs. By the end of the 1960s, there were more than 500 RCCs in business, and within a decade they and the Bell System were servicing nearly 1 million pagers across the country.[28]

In the 1970s the technology for person-to-person, anytime–anywhere communications began at last to surface. General Electric and a number of consumer groups pushed the government to allocate spectrum for a new personal radio service that would operate like a cordless home telephone, only with much greater range and no airtime charges. Unfortunately, neither GE nor the consumer groups knew how to accommodate growth on this system, except by adding new spectrum – the same old problem.[29]

The Bell System offered a different solution based on the work of D. H. Ring, a researcher at Bell Labs. Ring had first articulated the cellular concept in an unpublished paper soon after World War II. As early as 1949, AT&T had asked the FCC to provide enough spectrum to enable the firm to develop this innovation.[30] But the government, under pressure from television broadcasters, demurred. As a result of the usual combination of competitor and regulatory resistance to change, AT&T did not begin testing the commercial potential of Ring's cellular concept until 1962.[31] Using this approach, Bell divided the land into zones or "cells" ranging from one to twenty miles in diameter, depending on the terrain and demand within a particular area. Customers in a given cell would send and receive to a single transmitter or base station. Transmitters in adjacent cells would use different frequencies to avoid interfering with one another. Relatively low antennas and low-powered transmitters meant that cells that were sufficiently far apart could reuse frequencies – just as radio stations in different cities do. By "splitting" cells, or making them smaller, AT&T could increase the capacity of the system to serve more customers.[32]

The initial challenge to Bell technicians was to learn how to "hand off" calls from one base station to the next as a mobile customer passed from one cell into another. In the 1970s, advances in electronic switching systems – combined with low-cost frequency synthesizers and high-capacity microprocessors – at last enabled Bell researchers to make cellular work on experimental systems in Newark and Philadelphia.[33] The first cellular phone call was placed in January 1969 on a system built by AT&T for pay-phone customers aboard Amtrak's New York-to-Washington Metroliner.[34]

During the same period, Motorola was pushing to develop car phones and portables. Martin Cooper, a Motorola executive who would later be dubbed "the father of the cellular phone," placed the world's first call on a portable, hand-held cell phone from a street corner in Manhattan on April 3, 1973. The phone was an early model of "the brick" that Sam Ginn would use at the 1984 Olympics.[35]

AT&T and Motorola put up the first full-scale demonstration cellular systems in the United States in 1979. AT&T launched its system, known as Advanced Mobile Phone Service (AMPS), in Chicago.[36] When fully deployed, it covered 2,100 square miles and served approximately 2,000 customers, most of them selected at random from lists of businesses.[37] The system was wildly successful. Contacted by direct mail, telephone, and personal visits, one out of eight companies that were approached chose to participate. Construction companies, real-estate agents, and high-level salespersons were especially eager to try cellular, and they were delighted with what they bought. AT&T soon began receiving letters pleading to be included in the program. "I have been on your mobile telephone list for four years," the president of a manufacturing company wrote, "and would appreciate anything you can do to arrange this for me."[38] A real-estate agent begged to participate in the trial. "I think expanded mobile phone service is a great idea," she wrote, "and am only surprised that in this day and age it hasn't been made widely available at a reasonable cost before this."[39] Tom Ehlers, a chemical company executive, said he had business associates and contacts who would kill to have one of the new phones. "They were offering me money to get them on the list."[40]

The AMPS trial was successful in part because the Bell System, true to its heritage, had delivered a gold-plated Cadillac in terms of network quality and engineering. The 2,000 customers lucky enough to participate in the program enjoyed ample spectrum and high fidelity.[41] "With the windows rolled up," Tom Ehlers said, "people couldn't tell I was talking to them on a car phone."[42] Customers told AT&T that the new cell phones saved them time and increased their productivity by 20 to 30 percent. Without advertising the service, AT&T received hundreds of inquiries a month from people interested in signing up.[43]

The public clamor for cellular mobile phones reflected a technological triumph that was more than seven decades in the making. It also suggested that the Bell System had grossly underestimated the popularity of this new service. This was a pattern that would be repeated many times: the entrepreneurs – whether individuals or organizations with substantial experience in telecommunications – as well as federal and state regulators would repeatedly underestimate consumer interest.

Could the process of technical innovation in wireless have been accelerated? Probably not during the decades between Marconi's invention and World War II. The relatively slow technological advances of those years reflected neither regulatory conservatism nor corporate underestimates of demand, but rather technical problems that had to be solved in transmission and switching. But after the war, both governmental and corporate conservatism imposed significant constraints on innovation. As late as 1982, most of the public and private interests connected to the industry – including Phil Quigley, Sam Ginn, and most other Bell-heads – were continuing to underestimate the potential of wireless to provide a broad public benefit and meet the needs of a mass market. But some entrepreneurs out there had been reading the AMPS reports to the FCC, and one in particular could see that the world of communications was about to be transformed.

FOUR

Innovation amid Regulatory Chaos

Craig McCaw could appreciate the value of spectrum. The second of four sons of a flamboyant Seattle cable TV and radio promoter named Elroy McCaw, the young Craig grew up hearing the stories of the money people had made in the television industry when the government first gave away spectrum. Reserved but extremely bright, he attended the same private school in Seattle where Microsoft founders Bill Gates and Paul Allen were educated. From there he went to Stanford University. Then, at age 19, his life suddenly came apart. Home from his sophomore year at the university while his mother was away on a trip, McCaw found his father in bed one morning dead from a stroke. The family soon discovered that Elroy McCaw's legacy was a mass of debts. After the estate was settled – with attorney Bill Gates, Sr., representing the family – Craig and his brothers were left with a small, backwater cable business in Centralia, Washington, as their joint inheritance.

Without the shadow of his father's presence, Craig quickly emerged as the family's business leader. A brilliant strategist who was not afraid to take risks, a stickler for quality and customer service, and an eccentric manager who believed in empowering employees almost in the extreme, he began running the family business from his dorm room at Stanford. Over the next ten years McCaw increased the value of his family's cable company to about $200 million. Then, early in the 1970s, McCaw picked up a license to provide car telephone and paging services in Centralia and found himself in the polyester-suited world of the radio common carriers. Through the RCCs' trade association, McCaw heard about a new technology called cellular.

From the beginning, McCaw sensed that cellular could be a spectacular, once-in-a-lifetime opportunity. When he read that the FCC proposed to give each cellular licensee spectrum equal to the amount allowed for three and a half TV stations, he thought: if nothing else, this is a valuable asset. He also read AT&T's AMPS trial reports from Chicago in 1979. Based on his experience in paging and the car-phone business, he realized that cellular would be a winner even with a conservative estimate of demand. His gut instincts told him that the projections were far too low. As someone who hated the confinement of an office and loved to be on the move, McCaw could appreciate the idea that cell phones would have an enormous liberating effect. Prices for car phones and then portables were falling, and he decided that soon almost everyone would want one.[1]

From his experience in the cable industry, McCaw knew that the real trick to getting into the cellular business would be obtaining a license, or even a part of a license, from the FCC. Because the agency regulated both telephone service and the radio industry, it would be the battleground – and probably a bloody battleground – for the fight over cellular. McCaw's entrepreneurial projections were right on all counts.

∗

It had taken the FCC years to decide whether and how to allocate spectrum to cellular. Much of the delay resulted from ongoing battles between various constituencies, a pattern of regulatory combat that characterized the industry from Marconi to McCaw. The Bell System and the radio common carriers had waged war for years. The RCCs had won a major victory when they received about half of the available frequencies the FCC awarded in the 1940s, only to be outmaneuvered by AT&T, which blocked interconnection to the Bell System.[2] When the FCC finally got around to mandating interconnection, the Bell System charged more for the service than the RCCs could afford.[3]

Cellular didn't immediately change this situation, but it did for a time reverse the struggle, putting AT&T in the position of being thwarted by competitive resistance through the regulatory system. Aware at last that congestion on the frequencies assigned for mobile radio had become serious, the FCC in 1968 expressed interest in the development of a truly efficient high-capacity mobile telephone service and suggested that it would set aside UHF frequencies for this purpose.[4] Development of an experimental program, however, was slowed by continuing conflict between broadcasters interested in preserving UHF channels for television and advocates of mobile telephone service who wanted this additional spectrum.[5] Yet after nearly two decades, the broadcasters had not been able to make UHF television commercially successful, so the FCC agreed in May 1970 that inadequate spectrum allocation was retarding the growth of mobile communications. Alas, the Commission failed to decide how much spectrum to allocate and to whom. It did, however, authorize Bell Labs to develop and test the cellular concept under real-life conditions – the roads and highways of Newark and Philadelphia.[6] It also made a controversial decision about the future of competition in cellular.

Initially, the FCC planned to allow only AT&T and local exchange telephone companies the right to develop cellular.[7] The Commission, along with many people in the industry and government, envisioned cellular as little more than an improved version of the old mobile radio telephones, a local service. As AT&T pointed out, the Bell System had invented cellular and thus deserved the right to benefit from the innovation. Only the Bell System, AT&T maintained, was technically sophisticated enough to introduce cellular and sufficiently capitalized to pay for the new infrastructure. The RCCs' trade association agreed. They were content to pursue frequencies at a lower end of the spectrum that the FCC proposed to make available to supplement their existing systems.

Motorola, however, read with alarm the FCC's initial proposal to license cellular exclusively to Bell.[8] If cellular killed the mobile telephone and dispatching businesses of the RCCs – which seemed likely given Bell's resources and economies of scale – then Motorola would lose a substantial market. Not satisfied to be AT&T's captive equipment supplier, Motorola looked to its strong base of RCC customers for support; it encouraged them to fight the FCC and Bell for the opportunity to build independent cellular networks. Motorola also concluded that the best way to rebut AT&T's arguments for exclusivity was to show the FCC that someone else could design a better system. While AT&T was focused on meeting the need for car phones, Motorola concentrated on designing a system that would deliver anytime–anywhere communications. As one Motorola manager put it, "There are a hell of a lot more people than there are autos, and when you park your car and leave, you can't use your mobile radio, but you can take your portable [phone] with you."[9]

Protests to the FCC convinced the agency to reconsider its decision and face two key questions: How many cellular systems should it allow in each local market? And who should be allowed to compete for these licenses? The following summer, it reversed its decision and invited proposals from all comers for the development of efficient, high-capacity mobile systems.[10] Motorola jumped at this opportunity and submitted an alternative design for cellular to the FCC in December 1971. Meanwhile, the RCCs offered their own cellular alternative. Both of these submissions challenged the contention that only AT&T had the technical capability to develop an effective system.[11]

The FCC responded to these filings with ambivalence and equivocation. The government (especially the Department of Justice) was considering the introduction of competition in telecommunications, and it seemed hardly the time to give Bell another monopoly. That decision would put the RCCs and Motorola on the warpath[12] – as well as the Department of Justice, which urged the FCC to seize this unique opportunity to expand competition in telecommunications.[13] But the Commission flinched, still unable to make a decisive break with the past. In 1974, it finally decided to allocate a generous 40 MHz (megahertz) of spectrum to the development of cellular, but it returned to its original position that only the local wireline companies would be allowed in this market. Multiple systems in one market would, the FCC concluded, be too complex and costly.[14] Theodore Vail, the architect of AT&T's drive for universal monopoly service, couldn't have put it any better. The only exception to monopoly would be in the equipment sector of the new industry. By ruling that the wireline companies (and particularly AT&T) could not manufacture, provide, or maintain mobile equipment, the FCC bowed toward competition while dividing the opposition to its monopoly proposal.[15]

Once again, the agency's decision provoked a storm of protest. The RCCs and others took their case to the U.S. Court of Appeals. The Court, which expressed concerns about antitrust, nevertheless decided to withhold judgment on that aspect of the case and ruled in favor of the FCC.[16] In the meantime, the agency had already begun to backpedal. In the summer of 1975, it had opened the door for others to submit cellular proposals.[17]

Again Motorola raced to take advantage of this opportunity for innovation, applying for a license to operate an experimental cellular system in the Washington, DC, area.[18] The FCC finally gave the company's subsidiary, American Radio Telephone Service, permission to build the nation's second system in 1977.[19] This set the stage for competition between providers, technologies, and alternative visions of the future of the wireless world. It also set the stage for Craig McCaw's recognition that wireless presented an unusual situation that might lend itself to his style of entrepreneurship. Soon, the FCC's decision would bring Phil Quigley and Sam Ginn into wireless, promoting a different approach to innovation.

As expected, Motorola and the Bell System developed different concepts of cellular. AT&T continued to focus exclusively on car phones, and Motorola set out to create a system that would serve both automobile and hand-held portable phones.[20] Having licensed two experimental systems, the FCC decided to wait for the results before it made any further decisions.

When both AMPS and Motorola were successful, AT&T decided it would be unable to win the fight for a cellular monopoly. In competition with the RCCs for a single license in each market, the Bell System would probably end up with fewer than 25 percent of the licenses nationwide, making it very difficult to develop a coherent national service. Reading the tea leaves, AT&T wisely retreated and proposed a duopoly system with two licenses in each market.[21] The FCC accepted this compromise in a preliminary ruling issued on May 4, 1981, and – after weathering a flood of petitions for reconsideration and review – finalized the new rules.

Shortly after Charlie Brown and William Baxter announced their plan to break up the Bell System, the FCC laid out the details for the waiting entrepreneurs. To McCaw's disgust, the government was going to award licenses to two competing cellular providers in each metropolitan area. One would go to an existing wireline telephone company, which in most major urban markets meant a Bell System subsidiary like Pacific Telephone. From McCaw's perspective, these were all telecom Goliaths. The other license would be given to a "nonwireline" competitor, so McCaw would by design be David. This decision made McCaw and many others unhappy. State regulators complained that the federal government was preempting their historic jurisdiction over entry. The trust-busting lawyers in the Department of Justice asserted that the FCC's plan was "blatantly anticompetitive."[22] Most satisfied were many of the RCC crusaders. They had fought for more than a decade to be allowed to compete and were now happy to have a shot at half of the licenses.[23]

Soon the RCCs got a bitter lesson in how, exactly, the new duopoly structure of the industry would work. With the FCC's tacit approval, AT&T, GTE, and the other wireline companies met in 1982 and quickly negotiated a far-reaching agreement to allocate licenses.[24] The FCC approved this deal and thus, without extensive regulatory hearings to compare license applications, the wireline companies proceeded directly to "Go." The RCCs were furious. They had warned the FCC about the potential for a "head start" on the part of their well-financed competitors, but the FCC declined to take action. By late 1983, many wireline companies were already building their cellular systems and looking forward to profits from those high-rolling customers who had been on the waiting list for mobile service for years. In the meantime, the licensing process for the nonwireline companies became a twisted nightmare.

Initially, the radio common carriers tried to persuade the FCC to give existing local operators preference for licenses. But the FCC rejected this option and opened the competition to virtually anyone who asserted they could build a system. In keeping with the agency's historic pattern of allocating commercial frequencies, it proposed a series of "beauty contests" in which candidates would be judged by their engineering track records, financial capabilities, and specific proposals. The agency made it clear that it wanted sophisticated engineering studies.

On June 7, 1982, the day that applications were due for the nation's thirty largest metropolitan markets, teamsters in coveralls worked alongside well-dressed lawyers unloading trucks filled with paper. One company delivered its applications in an armored car. MCI organized thirty employees, each carrying a box of applications, to march from the company's headquarters in Washington, DC, to the FCC headquarters, six blocks away. One observer said they looked like explorers headed out on a safari. Graphic Scanning, Inc. – a company that had put the world on notice that it intended to conquer its side of the cellular business – sent a semitrailer filled with applications for all thirty markets. Meanwhile, Craig McCaw had teamed up with a handful of paging operators from the Pacific Northwest to apply for licenses in Portland and Seattle. He also put in applications on his own or with other partners for San Francisco, San Jose, Denver, and Kansas City.[25] But he was not in Washington that day, preferring to avoid the madhouse. Instead, one of his partners, Charlie Desmond (an executive with a Seattle-based paging company), had to step around the people still collating documents in the FCC's hallways to get to the counter and turn in the team's applications. By the end of the day, the FCC had received 194 applications, averaging well over 1,000 pages each, from companies that included MCI, Western Union, various broadcast and cable organizations like Metromedia, and dozens of RCCs.[26]

The prospect of sifting through these applications, holding comparative hearings for each market, and choosing among the competitors seemed overwhelming to some at the FCC. Over the next year, as deadlines for the second, third, and fourth market tiers rolled by, the number of applications swelled from the hundreds to the thousands. The Commission was swamped. The agency responded by compounding its errors. First, the regulators urged applicants to combine, taking shares in a partnership rather than going for sole ownership of a license. In an effort to streamline the process, the FCC apparently focused on those contestants proposing to provide the greatest coverage and then allocated licenses for the thirty largest markets in the nation.[27]

When the Department of Justice reasonably complained about this anti-competitive process, the Commission announced that it would resort to a lottery system for subsequent markets. This decision turned the launch of cellular into a nightmare. The number of applicants in later cycles ballooned as speculators decided to take a chance on the lottery. Why not? Fly-by-night scam artists promised in television commercials to write an application for an investor for $10,000, guaranteeing an 81.6-percent chance of winning a share in a license worth millions.[28] Many RCCs, which had worked hard for the right to compete, watched in shock as speculators "flipped" their licenses as soon as they were awarded. Some of the winners were political insiders who had seen these valuable license opportunities coming down the road. They included the first lady of Arkansas, Hillary Clinton, whose partners sold out quickly to Craig McCaw.[29] But others were Mom-and-Pop speculators who simply won the lottery. Rather than "regulatory failure," this was a study in "regulatory chaos." Sometimes, awards were rescinded, forcing the agency to repeat the process. Even when the agency appeared to have successfully awarded a license, the decision was often challenged in court.[30]

The outcome of this chaotic process was important. It would shape the future of the wireless industry and have a substantial influence on telecommunications around the world. While many of the speculators sold their licenses as quickly as possible, others – including Craig McCaw – were resolved to build new systems. With his first licenses in hand, McCaw set out to challenge the Baby Bells, all of whom he believed would be slow-moving and unable to keep up in the wireless race. He was almost right.

What McCaw could not anticipate was the extent to which Bell-heads like Sam Ginn could be transformed by competition into entrepreneurs – but not entrepreneurs like McCaw. When this painful and frequently erratic transformation was successful, it would produce innovators who would blend McCaw's style of rapid-fire, decentralized decisionmaking with the technological and administrative strengths of the Bell empire. For Sam Ginn, that process of change had actually begun several years before the wireless lotteries and the breakup, when he had arrived in Los Angeles to try to avert a crisis for AT&T.

FIVE

California Miseries

California is not Siberia, but to a rising star at AT&T in 1978, being sent to Los Angeles seemed like banishment. For years Pacific Telephone had been the black sheep of the Bell System. Its earnings were poor. Its managers held themselves apart from the rest of the System. The company was notorious for its high costs and low productivity. And its customers were just a little bit different. On the East Coast, at AT&T headquarters, some Bell System executives blamed California's laid-back culture and climate for Pacific Telephone's idiosyncrasies. Sam Ginn wasn't sure.

As he rode in a company car from LAX, he read the street signs. Names like La Cienega and Sepulveda made it clear that he was not in Alabama, Illinois, or even New Jersey anymore. But nothing on the street was as strange as the situation he encountered at the office. A dozen angry customers were waiting outside his door to complain about their service. When he looked into their problems, he discovered a nest of nasty issues. Here in the second largest city in the United States, Pacific Telephone had failed to upgrade its switching systems. To maximize the capacity of its existing machines, the company had developed the most elaborate trunking arrangement of any large metropolitan area in the country. This system, however, was hopelessly compromised by poor recordkeeping. When the manager in charge of the databases showed Ginn two sets of books – one cooked for AT&T, the other kept for internal consumption – it was like something straight out of a Raymond Chandler novel in which the good life under the southern California sun turns to evil and corruption at night. All that was missing was the corpse.[1]

Ginn quickly realized it would take months, maybe a year, to clean up the mess. Even if he fixed the engineering and network operations problems, he faced a greater challenge. Something *was* wrong with the culture of Pacific Telephone. Morale was low and the leadership seemed to be in disarray. At the heart of the problem lay a deep antagonism between California's regulators and AT&T. That antagonism threatened to develop into either a major service crisis or a financial breakdown. To understand the situation, much less fix it, one had to look back into the state's unusual history.[2]

✱

The roots of the conflict between AT&T and the California Public Utility Commission (CPUC) stretched back to the turn of the last century, when two interrelated events played a key role in shaping the character of utility regulation in California: the end of telephone competition in Los Angeles and the election of Hiram Johnson as governor. These events set the stage for a tug-of-war between the CPUC and AT&T that would last nearly seventy years – up to and beyond the breakup of the Bell System – and affect rates and policies in California and the nation as a whole.

At the turn of the last century, Los Angeles was the telephone capital of the nation. The city had two and a half times as many phones per person as Boston and four times that of New York or Chicago. By 1910, the city had surpassed Stockholm to become the telephone capital of the world.[3] Most New Yorkers would have been astonished. This was strange business for a community that Americans associated with orange groves and beach resorts.

Competition, together with the dynamic economy and population of Los Angeles, fueled the telephone's popularity in the first two decades of the twentieth century. But if Angelenos were eager consumers of telecommunications services, they were also vocal critics. They complained to city officials because Los Angeles had two competing systems, and subscribers to one service could not reach subscribers to the other. Businesses, particularly, had to install phones and lines for both companies, nearly doubling their expense. They lobbied the city to municipalize and consolidate the phone companies, or at least to force them to interconnect. This campaign put Los Angeles at the center of the nationwide struggle over the issue of competition and regulation in telecommunications – a struggle that eventually led to the Kingsbury Commitment in 1913.[4]

Los Angeles and California, however, were never quite comfortable with the Kingsbury compromise between monopoly and regulation. City officials debated competition, interconnection, and municipal ownership for four more years. When the city finally conceded regulatory authority to the CPUC (then known as the California Railroad Commission), it extracted concessions from AT&T in exchange for granting the Bell System an exclusive franchise. Under this agreement, AT&T's southern California subsidiary was not allowed to purchase equipment exclusively from Western Electric or enter into any agreement to pay AT&T's "license" fee of 4.5 percent of gross revenues. The city also retained the right to purchase the telephone system sometime in the future. Though Los Angeles never chose to exercise its option, the limits it placed on the phone company threatened the basic financial structure of the Bell System and set the stage for long-lasting tension between California and AT&T.[5]

The CPUC was one of the strongest state regulators in the nation.[6] It took a much tougher line with Pacific Telephone and the Bell System than regulators from other states.[7] In the 1930s, it favored consumers on issues related to depreciation.[8] During World War II, long-time CPUC head Ed McNaughton played a key role in developing a new national program to draw a subsidy from long-distance charges to help pay for the cost of local phone service.[9] McNaughton shaped the culture of the CPUC just as surely as Theodore Vail had shaped the culture of the Bell System. McNaughton, who had come to the commission in the 1920s, stayed for forty years. He believed the commission's job was to implement the legacy of Hiram Johnson, and that meant challenging the utilities at every turn.[10] If any part of Pacific Telephone or AT&T's business had escaped federal regulation, the CPUC should regulate it, he told staffers. As a powerful member of the National Association of Railroad and Utilities Commissioners (NARUC), McNaughton encouraged other state commissions to watch AT&T and the Bell Operating Companies with a critical eye.[11]

By the 1970s, the tensions between AT&T and the CPUC had produced a major crisis. For the average man on the street, the immediate conflict couldn't have been more confusing. A 1954 federal tax law had made it possible for utilities to enjoy tax savings by accelerating the depreciation of some of their assets. These tax savings were supposed to help companies build up capital reserves for reinvestment, thus providing a stimulus to the economy. But AT&T and its operating companies chose not to take advantage of the new law because they feared that regulators in California and elsewhere would force them to refund the savings to consumers.

AT&T's fears were well founded. In a series of decisions in the 1960s, the CPUC decided that – since Pacific Telephone *should* have adopted accelerated depreciation – the Commission would calculate the company's rate base "as if" it had. The rate decisions that followed dramatically lowered Pacific's revenue, squeezing both its available capital for investment in new telephone plant and its returns to its parent company.[12]

AT&T refused to accept this decision. In its own calculation of Pacific's earnings, it treated the company "as if" it used straight-line depreciation. AT&T also lobbied Congress to pass legislation that would prevent the CPUC from passing on the tax savings of accelerated depreciation to consumers.[13] When Governor Ronald Reagan's appointees to the CPUC finally sided with AT&T on the issue, the California Supreme Court intervened and ruled against the telephone company.[14]

The legal battle over depreciation created a financial crisis for Pacific Telephone. As California's population boomed, the company borrowed more and more money to keep up with the demand for telephone service. While the CPUC and the California Supreme Court were lobbing the tax issue back and forth, Pacific was receiving no rate increases. Moreover, California's decision to force Pacific Telephone to take accelerated depreciation exposed the company to an enormous federal tax liability that soon grew to almost $1 billion. The inflation rate was soaring, and the company's return on invested capital was sliding downward. Pacific consistently rated at the bottom of the Bell System in service quality and profits.

Starved for capital, Pacific Telephone couldn't invest in modern electronic switching equipment or adequately maintain its existing network. The quality of telephone service in the state continued to decline. Unable to provide first-class service like other Bell companies and oppressed by the intense antibusiness mood of the times, Pacific's employees and executives were demoralized. Defeatism engendered sloppy performances, and the Bell way of doing things was honored more in words than acts. To turn the company around, Pacific Telephone needed capital, commitment, and a new style of leadership.[15]

As Ginn studied this messy situation, he knew how important it was to avoid a major service crisis in Los Angeles. High-quality performance had, after all, long been AT&T's regulatory and legal card in the hole. People could complain about many aspects of the System, particularly its attitude toward customers, but it was widely acknowledged to be the best telephone service in the world. Each of the compromises AT&T had made with the government over the years had been facilitated by that fact. When there were glitches, like the well-publicized problems in New York City in 1969, the Bell System had responded by focusing enormous resources on the problem.[16]

Ginn responded in that tradition shortly after he began work in Los Angeles. As he told his New York bosses, there wouldn't be a quick fix. First, he had to conduct a thorough inventory of the L.A. system. Then he could start fixing the network incrementally. To do so, he needed help – lots of help. He asked for and received 400 temporary workers from Indiana, Ohio, and other states to help repair the L.A. network and restore it to System standards. Resources became all the more important to Ginn in late 1979, when he was promoted to executive vice-president (EVP) and given responsibility for the firm's entire network.

Unfortunately for Pacific's customers and its new EVP, the company's situation continued to deteriorate. Appeals to the U.S. Supreme Court to overturn the CPUC's decisions on the tax issues failed. An application for rate relief produced instead further rate reductions.[17] At Pacific's board meeting in August 1979, the company's general counsel said he expected a bill from the IRS soon. Desperate for capital, Pacific's financial officers turned again to the credit markets, only to find that the company was virtually tapped out on its borrowing capacity. With AT&T still refusing to contribute any equity, Pacific Telephone was on course for a catastrophe.

Instead of capital, AT&T provided tough love and another new leader. As president of Illinois Bell and later as chief financial officer (CFO) of AT&T, Charlie Brown had long been a fierce critic of Pacific and the main architect of AT&T's strategy of withholding equity until the CPUC granted a significant rate hike. As CEO of AT&T, he dispatched a new leader to turn Pacific around.

Don Guinn, who was a pragmatic and dedicated Bell-head, became Sam Ginn's new boss and mentor. A 26-year veteran of the Bell System, Guinn had grown up, like Ginn, among the working poor. Born in Wellington, Kansas, during the Great Depression, he moved with his mother (after his parents separated) to Portland, Oregon, when he was 7. At an early age, he took odd jobs in construction to support himself and bring money into the household. He put himself through college at Oregon State University in Corvallis, majoring in engineering; he was not in love with the idea of building things, but he believed that engineering would discipline his thinking and teach him how to solve problems. He was right. Graduated at the top of his class, he chose to work for the telephone company because he thought it would allow him to stay in Oregon. Clearly, he didn't know how the System trained its most promising managers.

Over the next twenty years, Guinn moved around the Pacific Northwest, went east to AT&T's headquarters, and then returned to Seattle for a series of engineering and network management jobs. By 1976, he had been promoted to AT&T's new headquarters in Basking Ridge, New Jersey, to lead the introduction of automated systems and the general modernization of the network. He liked this job and was unhappy when Charlie Brown gave him his new California assignment in 1980.

Hardened by now to the Bell way of doing things, Guinn quickly got over his own feelings and began to focus on applying his disciplined style of leadership to Pacific's deepening crisis. He started with a long series of interviews with AT&T officers and executives at Pacific, carefully jotting down notes in a small black book. Then, he held a three-day meeting at the Silverado Country Club in Napa in order to focus everyone's attention on the main problems and the strategies they needed to adopt if they were going to avert a major corporate meltdown. He had some of Charlie Brown's style of tough love to offer. He likened the management team, including Sam Ginn, to an alcoholic who cannot recover until he admits his problem. It was time, Guinn said, for all of them to face reality. He also said that he had confidence in his managerial team, but the "love" side was muted and the "tough" side was up front.

To turn itself around, Guinn said, Pacific had to get on top of its operations as well as its finances. This meant improving service as quickly as possible, the problem with which Sam Ginn had been struggling in Los Angeles. Fortunately, Ginn had already made progress on that front, starting Los Angeles on a course that would soon bring it up to Bell standards. With a good-faith show on operations, as Guinn explained at Napa, Pacific Telephone would be positioned to move forward decisively if it could resolve the depreciation tax issue with the federal government, win a major rate increase at the CPUC, and convince AT&T to contribute new equity. After the Napa meeting, Pacific's executives moved forward aggressively on all of these fronts. While Sam Ginn continued to focus on operations, Guinn went to work building the company's rapport with the CPUC, promising candor and information.

He found in CPUC Commissioner Richard Gravelle a receptive audience. Gravelle was worried about Pacific's future. A former staff attorney appointed to the agency by Jerry Brown, Gravelle recognized that the game of chicken being played by the Commission and AT&T could well end in disaster for California telephone customers. He held out an olive branch to an AT&T attorney who had once worked for Pacific.[18] Gravelle's overture resulted in a tacit peace plan between AT&T and the CPUC. The Commission would expedite consideration of Pacific's appeal for emergency rate relief and would not block any financing proposals generated by Pacific in the interim. AT&T agreed to make a limited equity infusion in 1980. The following spring, the CPUC kept its end of the bargain by granting a $227-million rate increase to Pacific, the largest single rate increase in the history of the Bell System.[19]

Welcome as the increase was, it was not enough. Pacific had been running on debt for too long. Even with the award, the company's credit rating did not improve. Earnings remained flat. Operations were stretched to capacity and the firm still desperately needed money for modernization. In August 1980, Pacific filed a new request, this time for a $795-million increase.[20] As Guinn told the commissioners during public hearings: "The string has run out." Without an increase in rates, Wall Street would no longer buy Pacific's bonds. Unable to sell stock or debt, Pacific would have to curtail expenditures – severely. The company would cut its projected $3-billion construction program substantially, which would inevitably generate a service crisis. Pacific would first take care of existing customers, leaving new customers waiting weeks or even months to get service.[21] California would resemble one of those foreign countries in which, Bell-heads like Sam Ginn said, "half of the people are waiting for a telephone and the other half are waiting for a dial tone."

Initially the Commission reacted negatively to Pacific's proposal,[22] but soon the effort to build a more trusting relationship between the firm and the agency paid off. On August 4, 1981, the CPUC gave Pacific Telephone a $610-million rate increase, authorizing an unprecedented 17.4-percent rate of return.[23] One of Pacific's three financial objectives had been met! They still needed equity from AT&T and relief from the tax issue, but at least Pacific's relationship with the commissioners at the CPUC had become, for the moment, cordial and constructive. Over the next several months, dramatic events would bring them into an even closer alliance.

*

Shortly after announcing that he had agreed to break up the Bell System, Charlie Brown sent a message to all of the officers of the Bell Operating Companies. The breakup, he said, would be the last heroic act of the Bell System. He wanted no squabbling over the family silver, particularly in Judge Greene's court. He expected teamwork, a minimum of internal dissension, and no public revelations of controversy.

There were hundreds of issues to be resolved, but for Pacific Telephone the greatest problems were financial. If the West Coast company was going to be viable after divestiture, it had to improve its balance sheet. Although the rating agencies put all of the Bell Operating Companies on a "credit watch," AT&T offered assurances that all the new Baby Bells would have adequate working capital for the next two years until the breakup was completed. Unfortunately, some public comments by top AT&T executives suggested that Ma Bell might divest the California company with a higher debt ratio and a weaker balance sheet than its six siblings. These remarks unsettled Guinn, Sam Ginn, and Pacific's other officers and board members.[24]

Fortunately, the firm's new ally, the CPUC, went to bat for Pacific in Judge Greene's court. The Department of Justice (DOJ) and AT&T had hoped to minimize interference from third parties, including the soon-to-be independent Bell Operating Companies, in their settlement agreement. But Judge Greene, who continued to be independent and a touch cranky, was of another mind. He wanted the public to have a voice in his courtroom. As a result, consumer advocates, industry groups, individuals, and competitors were able to raise issues related to the impact of the breakup on competition and consumers. State regulators also were able to speak up and express their opinions about the future prices of local telephone service and the financial health of the Operating Companies. Judge Greene asked AT&T to respond to these issues; when it did, it buried a landmine in one of its footnotes.

As AT&T carefully explained, it intended to divest all the Baby Bells with debt ratios of approximately 45 percent – except for Pacific. Pacific's debt ratio would be no better than it had been on January 7, the day the breakup was announced. At last, it seemed, the Bell System would make California responsible for the legal and regulatory chaos the state had created over the previous two decades. Given Pacific's history, AT&T said, the company had no right to expect anything better.[25]

Pacific Telephone's officers and Board of Directors did indeed expect something better – that is, equality – and became deeply concerned. At its May meeting, the transformation from loyal Bell-heads to independent telecom executives and board members accelerated. Pacific's Board instructed management "to seek an understanding with AT&T that at divestiture Pacific's debt ratio should be comparable" to the other Baby Bells. If AT&T failed to deliver, the Board said, then Don Guinn and the other officers should break ranks with the System by defying Charlie Brown's directive and going to Judge Greene. Nervous, but resolved to their task, Guinn and his team flew to New York to face men who had been their colleagues, mentors, and leaders for years.

The meeting in New York sparked bitter arguments between the two sides and produced an ambiguous, and later much-contested, result. Pacific's officers believed that AT&T made a commitment to bring Pacific's debt ratio down to 45 percent at divestiture – equal to the other Baby Bells. Based on that commitment, Pacific did not file a comment with Judge

Greene by the June 14 deadline. But in the weeks that followed, AT&T failed to produce a written document reasserting its verbal commitment. In fact, it continued to waffle. To Pacific's dismay, Judge Greene approved the modified final judgment (MFJ), with amendments in August, and accepted AT&T's plan to divest Pacific with a 50-percent debt ratio that would make it difficult for the firm to raise the capital it clearly needed for modernization. Pacific seemed to have lost this war.[26]

But thanks to Judge Greene's suspicions about AT&T and the determination of Pacific Telephone's executives, the company got to fight one more decisive battle. Judge Greene was deeply concerned that AT&T might leave the Baby Bells without the resources they needed to succeed. He ordered all of the CEO-designates of these new firms to file affidavits stating whether or not they believed their companies would be financially viable. This set the stage for a final struggle over debt.

Charlie Brown, determined to resolve the issues with Pacific Telephone, summoned Guinn and his team – which this time included Sam Ginn – to Basking Ridge, New Jersey, to talk until the issues were resolved. Ginn, now serving as EVP for strategic planning, was carrying the major responsibility for representing Pacific in the divestiture planning process. As he well understood, the meeting at Basking Ridge would establish the financial constraints that would control Pacific's (and his) strategic choices.

The two sides met for their battle at "the Pagoda," the nickname for AT&T's headquarters in suburban New Jersey. AT&T's officers lined up on one side of the table. Pacific filled the seats on the other. Sam Ginn and everyone else in the room could feel the incredible tension that developed. Dressed in a gray suit and with his hair slicked back, Charlie Brown looked especially intense. Ginn presented Pacific's list of operational issues. Some were complicated. Others were relatively simple, but basic. Whose operator, for example, should greet customers when they dialed "0"? Ginn marched them through a number of minor issues that were quickly resolved. But not the debt issue. By the end of the day, the group had made no progress on the fundamental question of Pacific's debt structure.

When it became clear that the debt question couldn't be resolved, Guinn played his strongest card: if AT&T didn't cut Pacific's debt, he would take his protest directly to the judge. Brown was furious, but he never lost his cool. As they parted, the two sides simply agreed to disagree. Pacific's officers flew home that afternoon; en route, Don Guinn and his general counsel, Robert Dalenberg, worked on the affidavit that Guinn would file with Judge Greene, the affidavit that would end Charlie Brown's dream of unanimity.

Pacific's decision to challenge AT&T in court brought a media spotlight on the tensions that the settlement had created within the System. Guinn asked Judge Greene to intervene, forcing AT&T to rewrite the Plan of Reorganization to correct the debt situation.[27] Characteristically, AT&T responded by blaming Pacific's problems on the CPUC. In his own affidavit to the court, Charlie Brown estimated that regulatory and judicial action in California had cost Pacific nearly $1.5 billion in earnings between 1972 and 1982. The company's earnings were so poor that it could not market equity and AT&T could not justify additional investments on its own. According to Brown, Pacific's financial health would be remarkably improved under the Plan of Reorganization. Any further improvement, he said, would be unfair to the shareholders of AT&T, the other operating companies, and the ratepayers in the rest of the nation. In essence, Brown said that Pacific's problems were California's problems and that California should deal with the consequences.[28]

Brown's public attack on the CPUC brought an instant reaction from the regulators in San Francisco. They countered that Brown was putting "the cart before the horse." The commission said that "AT&T's refusal to invest in Pacific from 1973 to 1980 was the major contributing factor to the poor financial condition of Pacific today." Moreover, in an earlier agreement with the CPUC, AT&T had agreed to reduce Pacific's debt ratio to 50 percent by February 1983. Now, the regulators told Judge Greene, Ma Bell was looking to escape that commitment through the process of divestiture. The CPUC asked Judge Greene to force AT&T to live up to its word.[29]

As Pacific prepared for its day in court, the company finally scored a victory that suggested the organization was starting to develop some of the capabilities it would need to be independent and more entrepreneurial. AT&T, which usually controlled all of the System's lobbying in Washington, had abandoned efforts to win legislative relief for Pacific's depreciation tax liability. That liability now stood at nearly $2 billion. Determined to fight, Pacific had opened its own lobbying operation in the nation's capitol. In a midnight meeting during the last hours of the 97th Congress, the company – in collaboration with several other California utilities that had been squeezed by the CPUC's actions on depreciation – won approval for a measure that would forgive the bulk of their outstanding obligations to the government. Pacific's CFO, John Hulse, kissed the company's top lobbyist, Art Latno, on both cheeks when he stepped off the plane in San Francisco.

For once, the regulators and the courts were all lined up on Pacific's side. With time running out, Charlie Brown began to fear that the mess in California would delay the scheduled breakup – the System's last great unified act. He ordered his financial people to settle.[30] AT&T agreed to improve Pacific's balance sheet by infusing $1.45 billion into the company.[31] Pacific's officers and board members were delighted. With these new financial resources, an improved relationship with the CPUC, and the freedom to launch new businesses, they thought their Baby Bell – Pacific Telesis – would be able to find rich opportunities for growth in the dynamic California economy. Sam Ginn, the firm's chief strategist of growth, was especially optimistic.

SIX

Can They Be Entrepreneurs?

Sam Ginn had never heard the word "entrepreneur" before 1968. It was not in the vocabulary of most of the farmers, factory workers, and small-town merchants he knew in Alabama. It was certainly not a word used in the Bell System. The culture of the Long Lines Division of AT&T was grounded in the values inherited from engineering and imposed by the elaborate state and federal regulations encasing every aspect of the network. The Bell System was highly creative, but its innovations all came from the top down, moved through bureaucratic channels, and were introduced by Bell-heads like Sam Ginn. Steeped in the System's internal controls, Bell managers played by the book – or they were soon gone.

As a Sloan Fellow at Stanford University in 1968, however, Ginn discovered a whole new way of looking at business, and he began to learn how to think outside of the Bell box.[1] The Alfred P. Sloan Foundation had established one-year education programs at Stanford and the Massachusetts Institute of Technology for mid-career, fast-track business executives. AT&T Long Lines annually picked two outstanding young managers to participate in the nine-month program and develop new managerial skills. This kind of academic rotation for mid-career executives in the Bell System traced its roots back to the 1920s, when Chester Barnard of Pennsylvania Bell established a program to send executives to the University of Pennsylvania to study liberal arts.[2] Barnard – who became a guru of Bell management – was institutionalizing AT&T's long-time interest in developing executives who combined technical ability with wide knowledge and experience.

Ginn thrived at Stanford. He learned how to dissect a balance sheet and an income statement. These were bottom-line skills that were rarely important to Bell managers, who were measured by service performance. He delved into case studies and analyzed organizational behavior in new ways. He focused intensely on how to motivate people, discovering the nascent Silicon Valley philosophy of giving employees stock options and thus a stake in the company's performance. As his major research project, Ginn looked at the effect of computers on the centralization of management hierarchies. He got a taste for academic life and liked it. The director of the Sloan Program encouraged him to stay and earn a Ph.D., but Ginn knew he was not inclined to spend the rest of his career studying the actions of others. Stanford whetted his appetite to manage and maybe, someday, to get a taste of entrepreneurship.[3]

Judge Harold Greene dropped that opportunity in Ginn's lap. Immediately following the settlement in the antitrust case, Wall Street predicted a dismal future for the Baby Bells, particularly Pacific Telephone. A *Fortune* cartoon depicted Pacific as an old beggar. In an era prior to the explosion of fax machines and the Internet, Wall Street's pessimism was understandable. The growth of basic telephone service had been slowing for years. As the industry's technology improved, many of its biggest customers had begun bypassing the local exchanges with their own private systems. The Baby Bells had reason to be anxious about their futures, and consumer groups and regulators echoed these concerns to Judge Greene and Congress. Greene listened.

The Judge decided to let the Baby Bells keep some of the profit centers that would subsidize the cost of basic residential telephone service. He undercut William Baxter's "elegant solution" and pressured AT&T and the Department of Justice to allow the Bells to compete in a number of traditional and nontraditional lines of business. Executives at Pacific Telesis and the other Bells looked at these opportunities to diversify with relief and delight.

As the man in charge of PacTel's strategic planning, Sam Ginn had to define the firm's diversification plan and then convince Don Guinn and the other officers that they should embrace the risks involved. Ginn and the team he collected were convinced that the company had almost no choice but to diversify. Divestiture had narrowed the scope of the company's business considerably. It now had fewer products and services to sell. As the smallest of the Baby Bells, it also had the shallowest pool of managerial talent and the most problems gaining access to capital. Confined to a single region of the country, Pacific Telesis was most vulnerable to local recessions. With 99 percent of its revenues derived from California, the company was hostage to the decisions of the CPUC. Given the

Cartoon from *Fortune* magazine (June 27, 1983) – On the eve of divestiture, *Fortune* magazine depicted Pacific Telesis Group as the most derelict of the seven Baby Bells. The company desperately needed capital to modernize its network.

history of Bell–CPUC combat, it didn't seem prudent to Ginn's team to leave shareholders in such a vulnerable position.

But clearly diversification would be an extremely delicate issue for the regulators to handle. Neither the CPUC nor any other regulatory agency had been created to oversee entrepreneurial ventures. By law the agencies were concerned with prices, profits, and service in specific markets. They were not charged with responsibility for innovation, and that was Sam

Ginn's worry. Over the years, the CPUC had been deeply concerned about the Bell System's potential to cross-subsidize or inflate revenues in one subsidiary with resources from the local telephone monopoly. Diversification under the flagship of Pacific Telesis would raise all of these issues again.[4]

Determined nevertheless to move ahead quickly, Ginn and his fellow executives worked out a two-pronged strategy. One line of attack would involve the "birthrights." Pacific Telesis and all the other Baby Bells were inheriting from the System some strong business opportunities – including Yellow Pages publishing and equipment sales. These "birthright" businesses would be the first out of the gate at divestiture and would most likely capture the attention of analysts and investors. All of the Baby Bells made plans to launch separate subsidiaries in these industries.[5] Further down the road, Ginn expected to launch a second diversification plan involving some "greenfield" startups that would attempt to leverage the telephone company's core capabilities and reputation. Real estate, equipment leasing, retail computer sales, and international telecommunications consulting all seemed to have some potential. According to the terms of the Consent Decree, however, Pacific Telesis and all of the Baby Bells would need specific permission from Judge Greene to enter these industries. No one was sure how long it would take to secure his approval.

Ginn was even more worried about how the CPUC would react to this strategy. Even if diversification were successful, he thought, shareholders might well lose out to ratepayers. The CPUC might try to use the earnings from these new businesses to reduce the rates for Pacific Telephone – now renamed Pacific Bell. After all, regulators had subsidized the cost of residential telephone service for years with revenues generated by long distance, and California regulators had led the national campaign to adopt this policy. Ginn mulled these complexities as he combed through the ranks of the company's managers looking for Bell-heads who had the potential to become entrepreneurs.

By the fall of 1983 he had learned a great deal about the Bell System. He knew what it could and couldn't do. He understood that it was weak in motivating its employees – except during times of crisis, when floods, fires, or hurricanes disrupted service. Normally, there was too much hierarchy and too little delegation of responsibility and authority for decisions to be made quickly. To be truly entrepreneurial, Pacific Telesis would have to shed some of the bureaucratic rules that Ginn and his colleagues had internalized from the Bell book. Ginn hoped they could do that without losing the technical capability that had long distinguished the System. But it was not obvious how they would be able to blend these two ways of doing business. It was a risky venture, one that was likely to be criticized by the conservative shareholders that Pacific Telesis would inherit from AT&T and resented by employees in the phone company. Still, Ginn chose to blaze a new path rather than follow the old Bell way. Although most of the Baby Bells chose a similar path, not all were equally aggressive. Ameritech, for example, which had headquarters in Chicago and operations in the Midwest, chose a much more conservative strategy.

In launching what became known as "the PacTel Companies," Ginn deliberately set out to create a different culture. He established them as separate subsidiaries with their own boards of directors. He moved away from the command-and-control style of the Bell System and embraced some of the lessons on entrepreneurship he had learned at Stanford. The president of each of these new companies was given a long, loose leash and told to run his own organization with a minimum amount of oversight. There was, however, one inflexible rule: meet your business plan objectives annually and be profitable within three years or you're out of the game.[6] With this setup, Ginn hoped to find out whether his crop of Bell-heads could become entrepreneurs. He thought the executives running "greenfield" startups were likely to have a considerable amount of trouble being profitable within three years. He gave those managing "birthright" businesses better odds, if only because they already had a track record. One, however, was brand new, and even though his old mentor, Charlie Brown, had said the prospects for cellular were "spectacular," Ginn wanted to see the numbers before he became too enthusiastic.

✻

Very few people actually knew what cellular was in 1982, and many of them were uncertain about where it would be located after the breakup of the Bell System. But AT&T's CEO was clear on this issue. Only a few weeks after he announced that AT&T would divest the Baby Bells, Brown found himself in front of a Congressional committee explaining the logic of the divestiture. Representatives, who were already hearing from consumer groups and constituents back home, peppered Brown with questions about how the end of the System would affect the quality and cost of local phone service. Representative Tauke captured the mind of many people in the room when he asked if it was true that the Baby Bells were being divested as "sitting ducks because AT&T or other companies could come in and chip away at their local base."

AT&T's chairman strongly denied this assertion. "The thought that they will be restricted from offering new exchange services is just not so," said Brown. "For example, a service right on the edge of being offered by these companies will be a spectacular service, that is, cellular radio."[7] Although Brown called cellular's potential "spectacular," he actually had no idea of the real potential. One now famous study, conducted for the Bell System by the consulting firm of McKinsey & Company, predicted that by the year 2000 there would be 900,000 cellular customers in the United States. Motorola's analysts thought that figure, which was six times the number of mobile radiotelephone customers in 1981, was too high. But as events unfolded, the consultant's predictions turned out to be far too conservative – off by a hundredfold.[8]

Sam Ginn had no more precise estimates for cellular than Brown did, but Ginn did have complete confidence in the executive charged with setting up PacTel's wireless operation. Phil Quigley had been sent to New York in late 1982 to learn about cellular and AT&T's plans to roll out the new technology. He had returned to California in March 1983 to oversee the construction of the AMPS system and to launch this new enterprise.[9] One of the first things he did was establish the headquarters for the cellular company in Costa Mesa in southern California, close to the company's premier market – Los Angeles – and far from Pacific Telesis headquarters in San Francisco. Through the rest of the year, all of his preparations were made under the aegis of AT&T, but Ma Bell would never profit from his work. After divestiture, the AMPS division in California became a subsidiary of Pacific Telesis and one of the new PacTel Companies.[10]

The subsidiary – PacTel Cellular, or just PacTel – began constructing the Los Angeles system in the summer of 1983, racing against the calendar. Their goal of having the system up in time for the Summer Olympics in 1984 seemed for a time to be impossible to achieve. With divestiture looming, AT&T had cut back on the capital allocated for cellular construction. Quigley thought these cutbacks were jeopardizing the work in Los Angeles and protested. AT&T, mindful that it was a major sponsor of the Olympic games, responded by increasing the capital budget for southern California, giving Quigley the funds he needed.[11]

PacTel's drive to get Los Angeles running in time for the Olympics was fueled by competition. Even before the network had been built, PacTel had lost a premier customer. The L.A. Olympic Organizing Committee had selected the nonwireline competitor in Los Angeles to be the official supplier of cellular telephone service to the 1984 Summer Games. But of course that firm also had to rush to get its network built, and regulatory battles over the license award had prevented it from pouring even the first concrete footing for its towers.[12] PacTel had an opportunity to steal the show, and both Quigley and his boss, Sam Ginn, were eager to seize the day.

Quigley, who had bragged that Los Angeles "was going to be the mega-market of the whole country," had something to prove.[13] As he knew, people in Southern California liked mobility and were early adopters of technology. But in order to prove his point, PacTel had to get its network in operation. Helped by the injection of additional capital, they succeeded, and on September 29, 1983, Quigley placed the first test call on the L.A. system. In January the company filed for final approval from the FCC and planned to be signing up customers within months.

To market its services, PacTel Cellular developed an extensive network of resellers who were responsible for signing up customers and selling or leasing them phones. This approach to marketing accomplished two things. It made the company's dominant role in this new service more politically palatable by sharing some of the wealth. It also relieved some of the pressure to invest in a system of stores and retail outlets, allowing PacTel to focus on what it knew best – building and operating a telecommunications network. Within months PacTel had signed up hundreds of potential resellers who began courting customers. Then, suddenly, the drive to introduce cellular quickly stopped. Once again, innovation yielded to the process of regulatory maneuvering and debate.

In Washington, the Bell competitors in the duopolies pressured the FCC to prevent the Baby Bells and other local exchange companies from getting a head start in cellular.[14] PacTel's competitor in Los Angeles didn't want Quigley's organization to get to market first and line up all the best customers. Unable to get their system in operation, they used the regulatory agency in an effort to slow the process of innovation and enhance their competitive position. They petitioned the FCC and the CPUC in 1983 to delay PacTel's initial construction permit and the start of service. Quigley and his team had used the Olympics as a high-priority public concern to pressure the CPUC and the FCC to let them begin construction.[15] Now, to persuade these potential competitors to withdraw their petition, Pacific Telesis had to compromise. PacTel agreed to provide interconnection guarantees and even to allow the nonwireline consortium to sell service on the PacTel network until its own system was up and running. This satisfied the FCC, which authorized PacTel to begin offering service on June 12.

For Ginn and his colleagues, the official launch of the company's first cellular system was auspicious, if somewhat comical. Los Angeles Mayor Tom Bradley tried to place the inaugural call to the U.S. Olympic Torch Relay team in Dallas, using a phone in the Cadillac provided for this dramatic occasion. But the battery was dead. Then he tried the phone in a Ford Mustang, but there was a bad connection. Finally, the call went through to an automobile in the Dallas–Fort Worth area that was driving alongside the torch runner. The runner stopped to take the call, Olympic flame in one hand and cell phone in the other.[16]

After this near debacle, the news for Ginn's cellular venture quickly improved. Demand in Los Angeles exploded, just as Quigley had predicted. Despite the high cost of the service – $45 a month, plus 45 cents a minute during peak hours and $2,500 to $3,000 for the phone itself – eager customers hurried to get attached to the network. PacTel's distributors and resellers had been installing phones in cars for months before the FCC at last issued its final approval. When the government flashed the green light, demand surged.

Ginn had told the Board of Pacific Telesis Group a year earlier that cellular was the opportunity of the century. Quigley was already on record with his optimistic prediction. Both men, like AT&T CEO Charlie Brown, thought they knew the meaning of "spectacular," and they were sure that cellular would be a spectacular business. But no one was prepared for the actual response. From the first day the service was turned on, a flood of new orders came pouring in. PacTel had projected 8,000 customers by the end of the year and 56,000 within five years. But by late July, less than two months after they had turned on the network, PacTel already had 7,000 customers and new subscribers were signing up every day. Ginn and Quigley were not alone in underestimating the interest in wireless. No one in the industry had predicted this kind of response and, for the next ten years, everyone running a cellular operation would consistently underestimate the potential demand.[17]

After his epiphany at the Summer Olympics, Sam Ginn was decidedly more optimistic, but he and his colleagues were just beginning to shed the caution imbued by the Bell culture. When they met as the Summer Games were ending, they carefully studied PacTel's subscription numbers. "Is this just Los Angeles?" Ginn wondered. "Will demand be as great in other places without L.A.'s car culture and freeways?" The consensus among the group was "yes," or at least "maybe." Either way, the message was clear. PacTel Cellular was a gold mine if they could keep the service humming and capitalize on some of the opportunities for growth.

They had a first-mover advantage that they could not afford to lose. Their nonwireline competitors were still struggling to get their licenses and build systems. Some of the other Baby Bells, including U.S. West, seemed to have decided that cellular would never be more than a niche business. This left an investment window open for PacTel. If Ginn and his team were right and demand was bigger than anyone seemed to be predicting, then cellular licenses were worth more than people were paying. By capitalizing on its experience, technical capability, and estimates of the cellular market's potential, the company would be able to turn those underpriced licenses into the kind of profitable business that every entrepreneur dreams about.

But they would have to move fast, and, as they recognized, the situation in Washington, DC, was going to make decisive action difficult – if not impossible. The nationwide rollout of cellular services had descended into chaos by the fall of 1984. Overwhelmed by the volume of applications for licenses, the FCC announced in February 1984 that it was changing the rules of the game. No longer would the agency hold comparative hearings or "beauty contests." Instead of awarding licenses on the basis of an applicant's track record and capacity to deploy and operate a cellular system, the FCC would use a lottery. As we noted before, this started a land rush to get in the lottery. Hundreds rushed to put in an application, regardless of whether they knew anything about radio or telecommunications. The confusion was multiplied by the battles taking place over licenses already awarded and the ongoing efforts to use the regulatory setting to prevent a competitor from getting a first-mover advantage. The chaos at the FCC left customers sitting on the sidelines, many of them with expensive cell phones linked to networks that companies weren't allowed to turn on.

The nonwireline companies had serious problems. It cost as much as $1 million per cell site in some locations to get a network running. They were left scrambling to form partnerships, consolidate their applications, and raise the capital they needed to move forward.[18] Some, discouraged by the problems of launching a cellular network, struck awkward deals with the wireline operators that allowed them to operate as resellers. That continued to be the situation in Los Angeles, where customers of L.A. Cellular were using PacTel's network.

Sam Ginn was impatient with the chaos created by the FCC, but he recognized that PacTel and the other Baby Bells were in strong positions in their wireless markets. Certainly that was the case with his cellular organization. The delay wasn't entirely negative, as it gave them time to build relationships with the agents and resellers who were then leading the sales effort.[19] They also took advantage of the opportunity they had to establish their own reselling subsidiaries in other markets. PacTel, for example, opened offices in New York, Dallas, and several other cities to get a feel for how those markets would develop and to keep an ear to the ground for possible acquisitions. Ginn and Quigley were thinking big already, long before Judge Greene had decided that the Baby Bells should become national players in a dynamic industry like cellular communications. But to accomplish big things for PacTel, Ginn and Quigley realized that they needed to pump more capital into cellular as soon as possible. Fortunately, the planets of political economy were aligned at that moment, and PacTel was able to fund decisive entrepreneurial ventures.

As Pacific Telesis came out of divestiture, its four-year effort to turn around the company began to pay off in a big way. Under Don Guinn's leadership, operating efficiencies in the local exchange company had improved dramatically. Regulators' concerns about protecting the profitability of local telephone operations had led to strong revenue streams, and the California economy had bounced back from a recession and rocketed upward in 1984, growing by 7.7 percent.[20] Finally, interest rates had fallen substantially from their 1980 highs, cutting the cost of Pacific Telesis's debt. All of these factors combined to make 1984 a very profitable year for Pacific Telesis shareholders and to provide Ginn with the capital he needed to pursue diversification more aggressively.[21]

Ginn and Quigley began to talk seriously about acquisitions, but as long-time Bell-heads they had never been swimming in the shark-infested markets for corporate control. Unlike other Baby Bells, Pacific Telesis had been uncertain in 1983 of the resources it would have available for diversification, so Ginn's early strategy was to experiment cautiously with startups. But after less than a year in the new world of telecommunications competition, he realized that acquisitions were essential if the diversification effort was to reach significant scale fast enough to make the company a major competitor in its new undertakings. That was especially true in cellular. To make deals, however, he needed people who could help identify targets, price the acquisitions, and nail down the contracts. Ginn needed new capabilities; fortunately, one of the people who showed up on his doorstep was a young, aggressive MBA who found the risky business of mergers and acquisitions exciting.

SEVEN

First Mover

Arun Sarin was pounding the pavement in San Francisco because his wife Rummi refused to move to Dallas. Sarin's family in India had been extremely wealthy and powerful during the British rule, but they had lost everything with the partition of the subcontinent in 1947. To feed their families and preserve a status that would allow their sisters to marry well, Sarin's father and uncles went into the army. Sarin's father, a colonel, sent his two sons to a military boarding school in Bangalore in the southern part of India.

Starting at the academy when he was 8, Sarin woke every morning six days a week to run five miles and work out in the gym before school. He applied the same kind of discipline to his studies. With a quick mind, an agile body, and an effervescent personality, Sarin excelled – in sports as well as academics. He was popular and, like his brother and his peers, imagined that he would become an officer in India's armed forces. But Sarin's mother had other plans.

A college-educated woman who could not bear the idea that she might lose both her sons in a war with Pakistan or China, she told her youngest son that he could not enter the Indian army or the air force. At the tender age of 15, he had to find another career worthy of his family's background. He contemplated the foreign service and imagined himself as a diplomat, but school officials encouraged him to take the entrance exams for the Indian Institutes of Technology (IIT). Good in math, Sarin scored well and was admitted to the IIT in Kharagpur near Calcutta. At his new school Sarin was among the top ten percent academically, but engineering was not his first love. He excelled in debate, drama, and sports, and he looked forward to studying management after finishing his five-year

Arun Sarin – A brash young MBA, Arun Sarin joined PacTel's wireless business in 1984 to target potential acquisitions. He became chief operating officer, and the heir apparent, at AirTouch in 1997. *Photo:* EricMillette.com.

technical program. In his final year at Kharagpur, he was awarded the B.C. Roy gold medal, which was given to the top all-around student. Prime Minister Indira Gandhi was on hand to present the prize, but Sarin was not there to receive it. He had already left to continue his studies in the United States.

The prospect of more technical education dismayed him, but the engineering program at the University of California, Berkeley, had offered him a full-ride scholarship. His friends and family insisted that it was an opportunity he should not pass up, and Sarin decided they were right. Moving into the International House in Berkeley, he made two crucial discoveries that reshaped his life. First, he met Rummi, who was getting an MBA in the Business School. Second, she convinced him to try a class in finance. Fascinated by the subject matter, Sarin received an A+ in the course and a suggestion from the professor that he ought to get an MBA. Confident in his ability to manage a greater workload, Sarin pursued both degrees at the same time and graduated in 1978 with an MS in engineering and an MBA.

At that point, he faced another difficult choice. After three years in Berkeley, one of the most hospitable environments in the United States for a foreign-born, razor-sharp man of color, Sarin did not want to return to India. Despite his family's background and his now prestigious education, he knew that at home he would start at the bottom in a bureaucratic society in which no young man could achieve a significant position in business or government before the age of 55. Sarin was impatient. Eager to move ahead, he saw that his business-school peers were getting great jobs right out of school. So instead of returning to India, he took a promising position with a management consulting firm in Washington, DC. He and Rummi married, and she went to work for Coopers & Lybrand's DC office.

Still impatient, Sarin moved again. This time he took a job in the Bay Area with Natomas, and Rummi transferred back to California. But when Texas-based Diamond Shamrock acquired Natomas and the company asked Sarin to move again, Rummi balked. "I've moved back and forth across the country already for you," she said, "this time you find a new job."[1] Fortunately for Sarin and his marriage, Sam Ginn was on the hunt for new talent to help with diversification. Hired with a number of others, Sarin was assigned to cellular by the luck of the draw.

Joining the corporate development team at PacTel, Sarin was in the vanguard of a wave of bright, well-trained Indian engineers, graduates of the IITs, who headed for California and Silicon Valley. They would play a critical role in the region's expansion, supplying much of the brainpower needed to fuel the growth of high-tech companies in the 1980s. In the next decade, many of them parlayed their experience and talent into entrepreneurship. They launched hundreds of companies of their own. Indian entrepreneurs and executives would run an estimated 10 to 15 percent of Silicon Valley's companies by 1999. As first-generation immigrants, the Indians were characteristically hard workers and risk-takers. As the cream of the Indian IITs, they were also very smart. These were the attributes Sam Ginn was looking for.[2]

✱

With money in their corporate pockets in the fall of 1984, Ginn and Phil Quigley were on the prowl. They asked Sarin to run the numbers and look for acquisitions that would be good values as well as tight strategic fits with PacTel's cellular operations. As Sarin surveyed the landscape, he looked for companies with sufficient scale to make a major difference in PacTel's market share and a significant contribution to the Pacific Telesis bottom line. Only three or four companies seemed to be likely targets, and most of them were paging companies that had won cellular licenses and were just beginning to build their networks.[3]

Sarin was not optimistic. The other Baby Bells and GTE on the hard-wired side of the duopoly were not likely to sell. On the other side of the market, there were loads of opportunities to acquire shares of licenses. In bars, restaurants, offices, and motel rooms, speculators, investors, and serious entrepreneurs were buying and selling shares in cellular licenses like they were playing "Monopoly." By the fall of 1984, a number of them had only just won their licenses and were in the first stages of planning their systems. Short on capital, some were already tired of the struggle. With market values for cellular rising, they began to look for an opportunity to cash in and get out, but neither Sarin nor Sam nor Phil Quigley knew if the FCC would allow PacTel to buy a share of a nonwireline license. After all, the Commission had made it clear that only half of the licenses were available to wireline competitors like Pacific Telesis.

In the absence of a rule about whether a Baby Bell could *buy* its way into those partnerships, Ginn decided to stick PacTel's nose into the tent in San Francisco. If he failed, PacTel would just have to go back to the other possibilities Sarin had identified. But if this maneuver succeeded, PacTel would open up some grand opportunities to use its capital and expertise to transform the structure of the U.S. wireless industry – and, not incidentally, to position itself as the first mover and market leader.

The choice of San Francisco for their test case was important to Ginn and Quigley. AT&T's earlier decision to let GTE take the lead in the San Francisco and San Jose markets had always rankled Ginn and other executives at Pacific Telesis. Under the original deal between the Bell System and GTE, PacTel had received majority ownership in the Los Angeles license while GTE received majority ownership in San Francisco.[4] But San Francisco was home to Pacific Telesis headquarters. It didn't seem right to Ginn and others that the company's cellular business would be invisible there, buried under the GTE brand. On the other side of the market (the nonwireline side) stood the Bay Area Cellular Telephone Co. (BACTC), which represented a consortium of companies – including McCaw Cellular and Dallas-based Communications Industries – with various stakes in the San Francisco and San Jose licenses. They all operated under the Cellular One brand name. One of the participants, Dallas-based Cellular Network, Inc., owned a 23.5-percent stake but wanted to cash out. When the other partners passed on the opportunity to buy Cellular Network's substantial share, Quigley told Ginn he wanted to offer $10.00 a POP (POPs are roughly equivalent to population or potential customers). The owners asked for $11.50, but Ginn told Quigley not to quibble. "Just go get it," he said.[5] When the deal was completed and announced, they waited for the sparks to fly.

They didn't have to wait very long. Craig McCaw was outraged. After winning licenses in Seattle and Portland in the FCC's initial allocation, McCaw had begun buying licenses or shares of licenses from anyone who wanted to sell. "We were actually the cleanup people to a defective application process," he said. "We just started vacuuming up the country, as much and as fast as our little hearts and wallets could allow."[6] The wallets were never big enough to match his ambition because he, unlike many others in the business, envisioned a nationwide system. Given his limited funds, however, he was forced to grow one market at a time.

His worst nightmare was that the Baby Bells would find a way to shut him out of the market for additional licenses. He could sleep easily with the duopoly because the FCC had, he thought, built a high wall that would keep the Bells out of his territory. Once he bought a license, he didn't mind competing with them because he believed they were sluggish and unresponsive to customers. The remnants of the old Bell System, McCaw thought, lacked entrepreneurial verve and the kind of competitive spirit he had. All he needed to best the Bells was capital – lots of capital – and a Federal Communications Commission willing to defend the duopoly it had created.

So when Sam Ginn and his cohorts suddenly announced in January 1985 that they intended to buy a share of the nonwireline business in San Francisco, McCaw exploded. From his point of view, the Baby Bells had been dealt all the good cards in the cellular game. They had been given their spectrum without having to compete or pay for their position in an open market. The FCC had provided them with a sweetheart deal and a head start in the business. Letting them into the nonwireline side of the market would, he thought, be an astonishing breach of faith by the government. PacTel's move was nothing short of an act of war, and McCaw and his team responded quickly and forcefully.[7]

They started by protesting PacTel's acquisition to anyone who would listen. In cooperation with other RCCs, McCaw asked the FCC to block the deal – the first of many, he was certain. He warned the Commission that PacTel would next try to force the nonwireline partners to use Pacific Bell's land-based links rather than microwave satellites.[8] He reiterated his argument that the Baby Bells would stifle innovation in the wireless industry. They had a tremendous investment in cables and wires, he said, and they would try to protect that capital. Consumers would suffer. Privately, he began to call Pacific Telesis "the death star."[9]

Although his protests had a great deal of validity, they fell on deaf ears in Washington. Ginn won the first battle in what PacTel and McCaw's teams later called their "Great War."[10] In March, the FCC approved PacTel's purchase as long as the company agreed to divest its 3-percent interest in GTE's San Francisco consortium.[11] McCaw was deeply suspicious of the agency's conclusion. He believed that PacTel had leveraged its close ties with the Reagan White House to influence the Commission's decision.[12] But the FCC seemed generally interested only in promoting competition in each market, even if the competition was between two of the Baby Bells.

Ginn tried to negotiate a truce with McCaw. PacTel, he said, was willing to compromise if they could establish a working relationship with McCaw in San Francisco. But McCaw remained belligerent. In response to one fax sent by Ginn, McCaw scrawled a two-word expletive across the top and faxed it back.[13] Ginn was appalled. These were not the kind of business manners you learned in the Bell System. But then this was no longer the protected world of rate-of-return regulation. For McCaw, The Great War had just begun and he was prepared to attack wherever his opponent seemed vulnerable – in the marketplace, the hearing rooms, or the courts. Ironically, PacTel's raid into McCaw's "territory" turned out to be an incredible boon to McCaw. Once the cash-rich Baby Bells were free to bid for nonwireline companies, the value of those firms, including McCaw Cellular, rose dramatically. McCaw was able to borrow more money and make more acquisitions.[14]

Ginn still felt that he now had the upper hand in The Great War, and Quigley and Arun Sarin renewed their search for potential acquisitions. No longer limited to one side of the duopoly, they soon found a decisive way to increase PacTel's market share. Armed with their firm's substantial retained earnings, they could think big and they did. In the spring of

1985, when an investment banker suggested that PacTel buy Communications Industries (CI), they were ready for a decisive move.

Communications Industries was an interesting company. Founded by two radiomen who had served in the Signal Corps in World War II, the business had merged in 1973 with Clayton Niles's General Communications Service. By the 1980s, the company was offering mobile telephone, pocket paging, telephone answering, and voice store-and-forward services in twelve states and the District of Columbia. To PacTel's officers, it was extremely important that CI's cellular subsidiary owned shares in licenses in Phoenix, St. Louis, Dallas–Fort Worth, Jacksonville, and Tampa–St. Petersburg, as well as a 23.5-percent interest in the Bay Area Cellular Telephone Co.[15] Communications Industries was a company with complementary assets and with the kind of scale that could make a major difference to Pacific Telesis's future.

After Arun Sarin had run the numbers and Quigley agreed on the target, Ginn took over. As a career Bell System manager, he had never been involved with mergers and acquisitions. He had never cut a high-level deal, so he was learning while doing. Flying to Dallas one weekend to begin his "lessons," Ginn met Niles for dinner and played tennis with him the next day. In between, they talked about their businesses and their vision for the future of cellular. Ginn made it clear that he was interested in buying CI if Niles was interested in selling.

As Ginn discovered, Niles was an entrepreneur with an unusually strong set of values. Ethical concerns had fueled his efforts to force the Bell System to provide interconnection to radio common carriers in the early 1960s. That struggle and his values had made him a leader in the industry. Now his ethics eased him toward a deal with PacTel. Niles knew that if he pumped in the money needed to develop CI's cellular licenses, he would hurt his company's earnings for some time to come. Quietly, he had begun talking to a number of deep-pocketed potential partners, including Bill McGowan at MCI, hoping they might be interested in capitalizing on CI's licenses. But he had gotten nowhere. With hostile corporate acquisitions increasingly dominating the headlines, Niles grew concerned that CI would become the target of a takeover. He decided that the best way to maximize the return to CI's shareholders was probably to sell the company.[16] Although CI was publicly traded, it still had many of the attributes of a family business, and Niles was clearly the patriarch. But he was also nearing retirement. If he was going to sell, he wanted to sell to a well-managed, well-financed company that would take care of his business and his employees. Trying to keep control of his own destiny, Niles hired investment banker and McKinsey veteran Barry Lewis to orchestrate an auction. He told Lewis to include BellSouth and PacTel as possible bidders.[17]

As Niles knew, a deal with Pacific Telesis would not be easy to complete. Communications Industries had manufacturing operations, and the AT&T antitrust settlement precluded the Baby Bells from manufacturing; hence this part of its business might have to be sold before any contract could be concluded. Communications Industries also owned a nonwireline license in San Diego, where PacTel was already competing on the wireline side. Again, one of those businesses would have to be divested. Ginn acknowledged these challenges but persuaded Niles that PacTel could solve the regulatory and judicial issues. He promised that CI's shareholders would be well compensated and worked hard to convince Niles that, at PacTel, CI's management team would have new opportunities to grow. Niles knew that this was a standard line for an acquiring company and that it was rarely acted on after the wedding festivities were over. Still, he trusted Ginn.[18]

As the talks turned serious, two potential rivals for CI's hand emerged. A group of investors led by a former executive of the company put together a leveraged buyout bid. But the offer was low, and Niles decided it was too risky for his shareholders. When the bids came in from BellSouth and PacTel, Ginn's offer was slightly higher, and Niles's board gave him the go-ahead to negotiate a final agreement. Ginn now had to move quickly. If BellSouth raised its offer, a bidding war would surely follow. In an effort to avoid that, he and his team flew to Dallas for a "lockup" negotiating session during which neither side would make or receive calls or other messages. With any break in the lockup, Clayton Niles would be forced to return the phone calls of BellSouth's CEO John Clendenin, so the two sides worked around the clock. The session lasted 36 hours! At one point, Ginn crawled under the table to sleep while the lawyers continued to work. Finally, the two sides smoothed out all the remaining details and finished the deal.[19]

PacTel agreed to buy Communications Industries for a whopping $431 million. The deal included several provisions that made it difficult for BellSouth or anyone else to intervene.[20] In the business press, a number of analysts expressed astonishment at the price Pacific Telesis was paying. They thought it reflected a far too optimistic view of the future of cellular. At times, Ginn and Sarin had similar doubts about this deal and others that they made. Their valuations were grounded in projected penetration rates based on their experience in Los Angeles, which might prove to be an unusual case. After less than a year in the business, they knew that penetration across the country was going to be higher than many people expected, but they were still guessing. They didn't know how high it would go. No one knew.[21]

Ginn now looked forward to further acquisitions, but Pacific Telesis was not prepared to crawl out any further on that limb. The conservative values of the old Bell System were still at work, shaping resistance to the kind of rapid expansion that acquisitions involved. Some in the Pacific Telesis organization, including CFO John Hulse, thought the company had paid too much for CI and for its stake in San Francisco. The company, they said, should take a breather. Like most of his peers in the Bell System, Hulse was not sympathetic to the 1980s style of corporate finance. He had arrived in his job as CFO at Pacific Telephone in 1980 after spending 22 years in a variety of different Bell disciplines, most of which focused on operations. Although he characterized himself as a big fan of cellular, he was convinced that share values were already too high by 1986.[22]

Arun Sarin – with his high energy, business-school training in finance, and favorable disposition toward mergers and acquisitions – offered a totally different outlook on the business. He had never really experienced the traditional Bell culture. He had never become a Bell-head, and in 1986 he argued forcefully that the best deals were available now. He and his colleagues in corporate development were convinced that they "were at an early stage of a revolution."[23] Supported by Sarin's analysis, Quigley also argued for further acquisitions. He was out in the marketplace beating the bushes for opportunities and finding plenty of prospective deals that he considered bargains. If the company waited, he said, more people would realize how valuable the cellular licenses were. The FCC had opened the way for all the Baby Bells to buy nonwireline properties, and if PacTel delayed, it would face intense competition from its siblings. The CI deal had already demonstrated how expensive that competition might become. Ginn, like Craig McCaw, was hoping that diversification efforts at most of the other companies would distract them from aggressive promotion of cellular. But both men knew they had to move fast.

The competition quickly intensified after the acquisition of Communications Industries. While PacTel slogged through the process of trying to get more than sixteen major regulatory approvals for its purchase of CI, a number of major deals slipped through Ginn's fingertips. First, BellSouth, smarting from its loss in the bidding for CI and rankled by PacTel's invasion of its markets in Atlanta and the rest of Georgia, paid $107.5 million for a 50-percent stake in Mobile Communications Corp. of America. This deal gave BellSouth a position in fourteen major cities, including PacTel's home, Los Angeles, as well as Houston, Indianapolis, and Milwaukee.[24] Bigger deals lay just around the corner.

In the late spring of 1986, Metromedia, the largest non-Bell wireless operator in the country, put its properties – including stakes in New York, Chicago, Boston, and Washington – up for sale. Seventy-one-year-old CEO John Kluge, a legend in the broadcast industry, had come late to the wireless party; but once there, he had quickly bought up paging companies and radio common carriers that had filed for or already won cellular licenses in 1983. These purchases were, however, expensive. The company's debt, which had risen to more than half a billion dollars by the end of 1983, equaled more than three and a half times the shareholders' equity.[25] Kluge's solution to this problem had been to take Metromedia private in a leveraged buyout and then sell the company's seven television stations for $2 billion to Rupert Murdoch.[26] Now Kluge had decided that cellular shares would soon flatten out. At his age he had no desire to stay in the business for the long haul, so he decided to sell.[27]

Sam Ginn wanted to buy. This deal would give PacTel a commanding position in the U.S. market, making it three times larger than any of its rivals.[28] Ginn and Quigley had dinner with Kluge and opened the subject of a possible deal. Kluge was receptive. But he had already worked out a price that he wanted, and it was considerably higher than the going rate for cellular. Ginn and Quigley agreed to get back to him. In California, they put Arun Sarin to work on the numbers, trying to come up with an offer. Kluge's initial proposal was high, but they were still confident they could close the deal. Then, before they could even make their offer, Zane Barnes, the CEO of Southwestern Bell, pre-empted them.

Coming out of divestiture, Southwestern Bell – with operations based in Texas, Oklahoma, Kansas, Missouri, and Arkansas – had suffered from weakness in its regional economy. In just two years, declines in real-estate values and a slumping oil industry left Southwestern in last place among the Baby Bells in sales growth and second-to-last in earnings growth.[29] Unlike most of the other Bells, Southwestern had not explored diversification. Analysts criticized CEO Zane Barnes, a 45-year veteran of the System, for his unimaginative leadership. But then Barnes, as if to answer the critics, stepped into the bidding for Metromedia. Moving quickly, Southwestern offered Kluge $1.65 billion for his wireless operations.[30]

Ginn was mystified when he got the news. Given 24 hours to match or exceed Southwestern's bid, he pulled together an emergency meeting with the Pacific Telesis senior management. Even though they were willing to analyze Southwestern's offer, they thought that Kluge might simply be bluffing. The bid was $350 to $400 million dollars more than what Pacific Telesis had decided Metromedia was worth. It effectively doubled the going rate for cellular franchises to nearly $50 for each potential customer (the POPs). "It's apparent that they did no due diligence on this offer," Ginn complained, obviously frustrated. But he didn't know for sure. After all, no one could be certain about cellular values.

The Metromedia deal caught Sam Ginn halfway between the cautious, methodical style of the Bell System and the entrepreneurial risk-taking style of the most innovative Information Age industries. When he tried to clear a counteroffer with his fellow Pacific Telesis executives, he quickly discovered that this time caution trumped vision. This was just not how things were done in the Bell book, and the rate-of-return culture was still vibrant at the top of Pacific Telesis. Don Guinn and John Hulse said the price was too high, and Ginn yielded, allowing Southwestern to buy Metromedia. With the advantage of hindsight, Ginn would characterize that lost opportunity as one of the biggest mistakes of his career.[31] He already knew that cellular was taking him into a business world that called for swift, decisive action. But long ago he had internalized the Bell values with their emphasis on systematic, low-risk development, and he was still finding it difficult to leave that culture entirely.

Ironically, Ginn had lost out to another Baby Bell. But as Southwestern's aggressive bid indicated, the Bells were not changing in uniform or entirely predictable ways. That was to be expected in companies that had inherited the Bell traditions yet were the vanguard of a transformation that would sweep through telecommunications around the globe. All of the government-owned telecommunications systems would face these same issues. Their managers would, like those of the Baby Bells, have fits of indecision and would find the risk-taking and competition of the new age difficult to master. Ginn and his Bell peers were, from this perspective, on the global industry's cutting edge, where the managerial transitions were breathtaking, essential, and frequently painful.

*

Despite losing out with Metromedia, Ginn, Quigley, and Sarin pushed ahead in their search for opportunities to expand PacTel's cellular footprint. All too often, McCaw was one step ahead of them. But not always. In late 1986 Quigley brushed shoulders with McCaw when PacTel was negotiating a deal to acquire five cellular properties in Detroit–Ann Arbor, Flint, Grand Rapids, Lansing, and Toledo. This time Ginn moved fast, very fast. He authorized a competitive bid of over $300 million within an hour, and PacTel was able to beat the agile McCaw at his own game.[32] This latest acquisition, which gave PacTel a presence in four of the nation's top ten markets, left it ranked third among U.S. cellular companies in terms of population coverage, but not subscribers.[33] By the time the two companies were fully integrated the following year, PacTel had accumulated more than 125,000 customers and was the world's largest provider of cellular service.[34]

But McCaw was still hot on PacTel's heels, ceaselessly complaining about the deep pockets of the Baby Bells. He helped fuel speculation that the Bells' real motive for acquiring cellular properties was to stifle the growth of the industry and close off a potential source of competition for their local exchange businesses. All the while, McCaw continued to be on the prowl. He bought cash-starved MCI's cellular operations in 1986 for $122 million, a sale that MCI would later deeply regret because it left the firm without a wireless foothold.[35] To finance that deal, McCaw turned to junk-bond king Michael Milken, who helped him raise the capital he needed to pay for MCI AirSignal and other acquisitions. McCaw was so highly leveraged that outsiders watched him as if they were looking at a man balanced on a high wire. In the middle of The Great War, Don Guinn told McCaw that his emerging empire was just "a house of cards." McCaw responded with a wry smile and kept on buying.

During these early years of the U.S. wireless industry, executives like Sam Ginn and Craig McCaw had to learn how to be competitors in one market and partners in another. It was not an easy lesson for U.S. business leaders to learn, and it was an especially difficult lesson for those who had been socialized in a highly regulated environment and were just learning how to be competitors. But new circumstances called for new approaches to business, as both McCaw and Ginn learned. Their Great War continued until late 1986, when they reached a tentative agreement to settle the antitrust suit over the San Francisco license, a disagreement that was at the heart of their conflict. Their new deal let PacTel manage the Bay Area company under the name Cellular One. Voting rights were split, but McCaw retained his infamous "shotgun provision": a brilliant contract clause that allowed him to set his own price for the San Francisco property in 1991 or force PacTel to sell. That clause notwithstanding, the truce held and led to further collaboration as the management teams discovered that they shared a common view of the future of wireless.

Both McCaw and Ginn believed that wireless was going to be bigger than anyone anticipated. By the end of 1986, the number of cellular customers nationwide had already climbed to nearly 682,000.[36] So impressive was the growth that McCaw decided in early 1987 to bet his future on this new business. Selling all of his cable assets, including the station in Centralia that had given him his start, he raised more money to buy wireless licenses and build new systems.[37] Later that year, when McCaw Cellular Communications went to Wall Street with its initial public offering (IPO), the company raised an enormous war chest of $2.39 billion.[38] Ginn was impressed. He offered congratulations to McCaw in a short note and complimented him and his team for doing "a superb job."[39]

As Ginn understood, PacTel was still constrained in ways that neither McCaw's enterprise nor some of the other Bells were. Ginn could not resort to junk bonds even if he wanted to. He had no major business to sell to raise cash. When Southwestern Bell bought Metromedia, the company financed the acquisition with debt, but Pacific Telesis officers and board members were reluctant to go in that direction. The history of the late 1970s and early 1980s was still fresh in their minds. Pacific Telesis had inherited millions of conservative investors who expected a triple-A credit rating and a dividend that kept rising year after year. Inside the business, John Hulse protected their interests. When PacTel made a play in the summer of 1987 for some South Florida cellular properties owned by the *Washington Post* and several other companies, Telesis rejected the deal. According to Hulse's group, the price (in the mid $40s per potential customer) was too high. A year later, McCaw paid $85 for the same deal.

The Pacific Telesis heritage worked against doing deals in other ways. Given the company's history, for example, executives had a hard time accepting the notion that they might be minority partners in any market. As a minority partner, they feared that the majority might decide to enter a line of business, such as long distance, from which the Baby Bells were banned by the Consent Decree. Then, PacTel would be forced to sell its interest at a distressed price. Several opportunities were missed because of these fears. Phil Quigley, for example, had the opportunity to buy a 25-percent stake in Hawaii for a reasonable price, but Pacific Telesis rejected the deal because the company would not have control.[40]

Ginn began to find all of this frustrating. Despite McCaw's spectacular initial public offering, Wall Street didn't seem willing to include the value of PacTel's cellular operations in the Pacific Telesis share price. The wireless business had no separate war chest equal to Craig McCaw's. One option that Ginn considered was to issue a tracking stock for wireless. Clearly, he thought, Wall Street's appetite for cellular was whetted. An initial public offering of 20 percent of the business would raise millions.

Ginn and Pacific Telesis decided to go for it, announcing the plan in September 1987. But their timing turned out to be disastrous. On October 19 the stock market plunged, wiping out billions of dollars of paper assets. In the ensuing months the market for cellular stocks recovered slowly but, in the meantime, Ginn and PacTel were forced to concentrate on growing the business from within.

EIGHT

Breaking with the Past

Shortly after the CI deal made headlines, Sam Ginn, Phil Quigley, and a team of Pacific's officers flew to Dallas. Their number-one priority was to encourage CI's executives and managers to stay with the company after PacTel took over. That was a tall order. They would have to integrate the personnel and cultures of two radically different organizations – without driving off the best people from either of the businesses. PacTel had cut its teeth on the biggest cellular market in the nation, Los Angeles, where rapid growth was both a blessing and a curse. Growth hid many management mistakes, but it also highlighted shortcomings. Communications Industries had grown up in the rough-and-tumble, intensely competitive paging business, where its success depended on its decentralized structure and obsessive attention to detail, costs, and profit margins. It needed PacTel's capital, but PacTel needed CI's ability to exercise close control of a decentralized organization. To get it, Ginn had to win the hearts and minds of CI management.

"Most of us were ready to bolt," Craig Farrill remembered. "We looked at PacTel as a huge monolithic bureaucracy – and from California. When you have Texans looking at Californians, right away you have a major problem."[1] A young radio engineer in 1980, Farrill had been hired by CI to oversee the construction of its cellular systems. He turned on the first of those networks in San Diego, in competition with PacTel, while the acquisition was pending legal and regulatory approvals. Farrill's misgivings turned to full-blown skepticism after PacTel brought the CI team to California to tour PacTel Cellular's operations.

As Craig Farrill could see, enormous demand had allowed PacTel to grow quickly without developing any of the managerial accounting tools necessary to monitor the business closely. He also became nervous when he looked at the engineering of PacTel's network. In a pattern characteristic of the other Baby Bells and incumbent telephone companies abroad, the phone company had built too few base stations and relied on high-power transmitters that tended to generate enormous amounts of interference. Some people in the industry blamed AT&T, saying the company had cut back on capital allocations for cellular in the year prior to divestiture since it knew it would never reap returns on that capital.[2] Whatever the cause, the results were limited capacity and poor call quality at certain times. Customers complained. In a letter to the editor of *Business Week*, one southern Californian wrote that 40 percent of the calls he placed were either dropped, interrupted by static, or confused with bits and pieces of another conversation. "I think the technology is only a step above Dixie cups and string, and it is considerably more expensive."[3]

The CI executives also expressed doubts regarding PacTel's interconnection with Pacific Bell, which was inefficient and poorly engineered – even in comparison with CI's own interconnection with Pacific Bell in San Diego.[4] PacTel didn't even know how much their type of interconnection was costing them, but as CI's officers knew, these operational problems had a significant impact on margins.[5] PacTel's overhead averaged around 19 percent of total cost; CI's cellular subsidiary ran at 9 percent.[6] PacTel had not developed central management information systems for tracking customer activity or inventory. Although PacTel knew how many customers it was signing up each month, it didn't know how many it was losing. The company was not tracking average revenue per subscriber per month. This was the kind of market-oriented data that was traditionally not important to the Bell System, where all customers were treated the same way regardless of the volume of business they generated.

Fortunately for Ginn, CI's executives were not afraid to speak their minds. They gave Ginn and Phil Quigley a specific proposal when the two men visited Dallas in the fall of 1985. Under their plan, CI's executives would take over the management of PacTel's cellular operations. Another CEO might have kicked them out the door for making such a proposal, but Ginn listened carefully. He didn't accept the plan but, after PacTel completed the acquisition in June 1986, he and Quigley reorganized the cellular business and gave major responsibilities to CI's managers – within both the cellular operation and Pacific Telesis. This was an amazingly bold stroke on the part of two Bell-heads.

To integrate the two organizations, they consolidated the headquarters of PacTel's cellular and paging company in Dallas at CI's main office.

Quigley even moved to Dallas for a year. They left CI's paging operation in the hands of CI's Charlie Jackson. Ginn delivered on his promise to Clayton Niles to give CI's managers the opportunity to grow within the PacTel organization by putting CI executives into key positions. Chief financial officer Mike Barnes became the CFO of PacTel Cellular. Barry Lewis, the executive in charge of cellular, took over corporate development for the combined company. Craig Farrill was in charge of engineering and operations. Ginn moved CI's president, Roger Linquist, to corporate headquarters in San Francisco to become the Baby Bell's treasurer. Many of these executives would play key roles in shaping the company's future, and they would ensure that CI values and operational practices would be dominant.[7]

The decisions Ginn and Quigley made at this time seem even more remarkable given that PacTel's cellular operations were significantly larger than CI's. Revenues in the L.A. market alone nearly equaled all of CI's cellular earnings in 1985. Regulatory delays had left PacTel unopposed in that vibrant market, a situation that would continue until March 1987.[8] But even though PacTel was large, Ginn and Quigley knew it was less efficient than CI, whose decentralized structure – particularly in its paging business – had bred a cadre of outstanding general managers. They did not want to lose that talent. Instead, they wanted to transplant to PacTel both the talent and the CI approach to business.[9]

Communications Industries' decentralized management structure had been one of the keys to the company's success in paging, and many people presumed that cellular would benefit from a similar approach. Ginn and Quigley had organized PacTel along these lines, but now – as they worked to integrate the two firms – their commitment to delegating control was tested. Remembering his days at Stanford, Ginn stuck to his plan. As they reorganized the company on the basis of individual markets, Arun Sarin played a key role in staffing each division with financial professionals. This strategy ran counter to the Bell tradition of rotating executives (who usually had engineering backgrounds) through the finance departments as they gained experience in all of the disciplines of the organization. But Sarin believed that his staffing strategy, which emphasized specialization and thorough professional training, would strengthen internal controls and make each market division more responsive to its bottom-line performance.[10] In these ways, PacTel's leaders were attempting to develop the organizational capabilities needed to be a major player in competitive markets.

By empowering CI's managers, embracing CI's decentralized structure, and emphasizing market-oriented financial controls, Sam Ginn was steadily drifting away from the Bell style of management. After promoting Arun Sarin to the position of chief financial officer of the combined company, they worked together to strengthen the business's financial controls. Now, they thought, they could keep competitors from cherry-picking their customers because of the quality or cost of the service. By building a superior organization and aggressively seeking acquisitions, they hoped to move PacTel ahead of the pack in the U.S. cellular market.

The move to Dallas marked a critical transformation for the cellular enterprise and for Sam Ginn. Many of his competitors, including Craig McCaw, were still counting on the bureaucratic heritage of the Baby Bells to keep them from becoming formidable competitors. Ginn was determined not to let that happen. He wanted to keep the service ethic and technical capability of the old Bell System while making his organization responsive to customers and to costs. With the acquisition of CI's licenses in Atlanta, PacTel became the first Baby Bell to move into a sibling's cellular territory. Ginn was convinced he could beat BellSouth in its own backyard. "He was a canny player," McCaw admitted, "and probably the only guy in the [old Bell] phone business that was."[11]

McCaw was right. Three years after its launch, PacTel's cellular business was growing fast. By the end of 1986 it had 95,000 customers, or nearly 14 percent of the U.S. market. The federal government estimated that, over the next three years, the number of customers across the country would more than double to 1.5 million.[12] Ginn and his colleagues thought those figures were conservative, and they planned to capture an increasing share of the growing U.S. market.[13]

Most of PacTel's customers, like cellular users nationwide, owned car phones rather than portables. As the average cost of installed car phones (with their "pigtail" antennas in the back) dropped from $2,500 to $1,000, demand surged. Meanwhile, the cost of a portable phone remained over $3,000, beyond the reach of most people. Portables also suffered because most cellular networks, focused on serving vehicles on densely traveled highways, were not designed to support service in buildings or residential neighborhoods.[14]

In this early period of development, third-party resellers still played a critical role in the business. PacTel and other companies depended on equipment retailers to bring them new customers, and they paid a commission of $150 to $300 for each new subscriber. But as the nonwireline companies came on line and the Baby Bells faced competition, they began to focus more on direct sales. For PacTel Cellular this entailed a shift from being network operators to becoming effective marketers. This transition had to be managed while PacTel and the other Bells were still spending enormous amounts of capital to build out their networks and were operating in the red. None of the major cellular providers in the United States were making a profit in 1986, although some markets – Los Angeles, for example – were in the black. Wall Street analysts predicted that the industry as a whole would not begin making money until 1990.

In the meantime, the attention paid to car phones helped fuel the growth of the paging industry as a poor man's alternative to cellular and a solution to the desire for personal mobility. Paging proved tremendously popular with salesmen, tradesmen, and messengers as well as doctors, lawyers, and other professionals who were not always near a phone. In this wing of the industry, aggressive price competition had already begun, and with it came increasing consolidation. PageNet (run by former CI executive George M. Perrin) bought up smaller operators and companies to build a nationwide franchise. Several of the Baby Bells also pursued paging acquisitions aggressively.[15]

While the U.S. industry was changing rapidly and Sam Ginn was focusing most of his attention on the consolidation of PacTel's domestic business, he also began to look abroad for new opportunities. He and his colleagues speculated about the Asian markets, but none of them knew much about those economies. Sarin was familiar with India, but none of the other executives in cellular had even traveled to the Far East. As Don Sledge, one of Ginn's trusted advisers, put it: "We spent our summer vacations on a houseboat at Lake Shasta or in a cabin at Tahoe. Once in a while someone went to Europe."[16] This was the way most people in the Bell System lived.

That was all about to change, however, when they met a character who seemed to walk off the screen of a Hollywood detective story: a mysterious Chinese-American woman from Pasadena who claimed she could get PacTel into something big in mainland China. That got their blood going. The future of plain old telephone service (they called it POTS) in California was one thing, but international business and the vast China market – well, that was something else. It would surely make Wall Street take notice. For Sledge and PacTel, it began with a birthday party.

NINE

Innocents Abroad

Don Sledge tried to sleep on the way to China, but he couldn't get comfortable in the blue leather seat. Business class on the CAAC flight was only a little less crowded than coach. The food was just as bad. High over the Pacific Ocean, while his traveling companion Thomas Klitgaard slept soundly, Sledge watched two movies – including the brand-new blockbuster *Star Wars*. The battle between the freedom-loving forces of the Rebellion and the dark ruler of the Empire helped pass the time on the thirteen-hour flight. Still, he thought, this was a hell of a long way to travel for a birthday party.[1]

To Sledge, a Texan from Lubbock, the airport in Shanghai was not unlike the Star Wars spaceport on the planet Tatooine. Like Luke Skywalker, he marveled at the exotic scene. There were no creatures with tentacles, but the communist soldiers with their black automatic rifles made the place seem just as forbidding. Klitgaard spoke to the officials in Chinese, and Sledge presented his brand-new passport, but it still took an hour to clear customs. Even to a long-time Bell-head, this was an extraordinary bureaucracy.

Ignoring the delay, Sledge and Klitgaard pushed on, launching what would become a new phase of wireless entrepreneurship and a set of distinctly Third Industrial Revolution experiences. It was not just that the experiences would be global in scope – that had been true for many Second Industrial Revolution industries, including automobiles and electrical power. What would be different was the complex nature of the international involvement, the types of strategic alliances on which it depended, and the global market for corporate control that would dominate the process of merger and acquisition in this and other Information Age industries.

Learning how to compete in this new environment would not be easy. Success would be preceded by failure. Neither Don Sledge nor Klitgaard had any idea of the magnitude of change ahead of them. Arriving on the historical eve of this revolution, Sledge saw the trip as a conventional sales call – albeit an exotic one.

*

Outside the airport terminal, at midnight the July air was hot and humid. Escorted by Chinese Communist party officials in their blue Mao suits, the two men entered an official car waiting for them at the curb. As they raced along the roads into the center of Shanghai, Sledge suddenly realized the lights were not on. The driver told Klitgaard that he was saving the battery. Sledge, who was unable to speak the language, tried to explain that as long as the engine was running there was no drain on the battery, but he couldn't make himself understood. Through the window, he could see people sitting outside their houses to cool off and moving in the shadows along the road. He was sure the car would hit someone before they made it to the hotel.[2]

Miraculously, they arrived at the nine-story Jing An Guest House without killing a single pedestrian. Inside, the sleek wooden furniture, reminiscent of the styles of the 1940s and 1950s, made it seem to Sledge as if they had stepped back in time rather than flown halfway around the world. He was relieved to find that the room was air-conditioned. But not long after he went to bed, the air conditioner went off; like the driver, someone was saving electricity. Sledge opened the windows. Exhausted, he collapsed into sleep, but well before dawn the noise of people and cars in the street woke him again. This time, he could not get back to sleep. He was simply too excited. Lying in bed, an absurd John Denver refrain about "the breezes here in old Shanghai" kept spinning through his head.[3]

He realized that Sam Ginn had given him an incredible opportunity. Ginn liked Sledge. They had similar roots. Sledge had grown up in the South "without a pot to piss in or a window to throw it out of." Determined, like Ginn, to climb into the middle class, Sledge had spent seventeen years inching his way up through the Bell System. Unlike many of his fellow Bell-heads, however, the 43-year-old manager had an ambitious, risk-taking personality, and he wasn't afraid to take charge. A quick study, he could see the big picture in any business situation. Ginn appreciated what Sledge could do and brought him onto his staff in 1982 to help with divestiture, putting him in charge of the "choke list" issues for the showdown with AT&T. Later that year, Ginn pulled Sledge into diversification.

By default, Sledge became the guy on the team who ran the administrative process and tracked down every proposal that came in over the transom. That's how he had encountered the woman from Pasadena.[4]

An immigrant from the People's Republic of China, Ms. Lee published a Chinese-language technology magazine.[5] Her husband, a scientist, worked in a lab at Cal Tech. The two of them traveled frequently back and forth to the East. She had learned that Pacific Telephone was replacing all of its electromechanical central-office switches with digital equipment. Quick to recognize an opportunity, she got in touch with Sledge's staff. She told them she knew people in China who were decisionmakers in telecommunications. She said that Pacific could make a handsome profit by selling its used equipment to China, and she offered to help put together a deal.[6]

Sledge knew nothing about Ms. Lee or her contacts in China, but the opportunity she suggested was tantalizing. China was so huge, the population so enormous, that it represented a tremendous market. Americans had been having China Dreams for over a century: "If only we could sell one razor blade, or T-shirt, or pack of chewing gum to each person in China," the dreamers said, "we would be rich."

Sledge floated the concept up to Ginn, who took it to the Telesis Policy Group. To Sledge's surprise, both Ginn and Don Guinn were interested. Very interested, in fact. From Ginn's point of view, China seemed like a natural opportunity. With its headquarters in San Francisco, Pacific Telephone had a historic relationship with the Chinese-American community. The City of San Francisco and the San Francisco Chamber of Commerce were actively working to promote trade with the Chinese, especially with the sister city of Shanghai. Richard Nixon's visit in 1972 had opened the way. With the death of Mao Zedong in 1976, the ouster of his hand-picked successor in 1981, and the rise of the more capitalistically minded leader Deng Xiaoping, the country had actively begun to court foreign investment. Delegations of American business leaders, including at least one of Pacific's board members, had toured China.

Ginn himself had gone as part of a San Francisco Chamber of Commerce tour shortly after he and Ann came to California. Like many of these business leaders, he was awed by the size of the potential market. He, too, began to have China dreams. And China could be just the beginning. The Pacific Rim's economy was growing in 1983 at an astounding rate. Altogether, the Asian nations were expected to spend $50.8 billion on telecommunications products and services in the 1980s.[7] If only Pacific could get a piece of that business. But "what if" and "if only" were a long way from Shanghai.[8]

Ginn sent Sledge to Shanghai to get a feel for the bureaucracy and the potential for Pacific to do business on the mainland. The birthday party was Klitgaard's idea. An outside attorney for Pacific Telephone, Klitgaard had gold-plated credentials in China. He had taken time off from his law practice to study Chinese at the famous Defense Language Institute in Monterey. He had traveled to China a number of times representing San Francisco Mayor Dianne Feinstein. Through his work on behalf of the city and the chamber he had come to know officials at the Shanghai International Trust and Investment Corporation (SITICO), which represented the new, quasi-capitalist face of Deng Xiaoping's China. An economic development fund founded by the Shanghai Municipal Government to promote investment in the region, SITICO was raising millions of dollars for infrastructure and trade.[9] Celebrating its second anniversary in July 1983, SITICO was planning a big birthday bash. Klitgaard had suggested that Pacific Telephone ought to send someone to the party, and Ginn picked Sledge.[10]

That first morning in Shanghai, Sledge ate ham, eggs, and toast and drank coffee – a touch of American culture that made it seem like he was not so far from home after all. The birthday party took place in the gardens of the old French Consulate, where Mao used to swim in the pool when he came to Shanghai. As part of the week-long celebration, Sledge and Klitgaard toured the newly industrialized area of the city. Klitgaard introduced him to a number of people. There were speeches by foreign businessmen and dignitaries, including an executive from the Swedish telecommunications equipment giant Ericsson. At one point in his speech, the Ericsson executive referred to his company's forty years of experience selling equipment to the Chinese, and Sledge suddenly felt intimidated by all the old China hands around him.

Nevertheless, the more he learned about the Chinese telephone system, the more excited he became. The Chinese lagged the Bell System technologically by several decades. Throughout the nation there were only four telephones for every 1,000 people, and most of that equipment was in government offices. Shanghai, the fifth largest urban area in the world and with a population estimated at more than 12 million, had barely one telephone line for every 1,000 persons.[11] What an opportunity! The government intended to more than double that capacity by 1990, adding nearly a million lines by the year 2000.[12] What a coup it would be if Pacific could provide the switching equipment for that service and turn a profit on its outmoded machines. There seemed to be even bigger opportunities to provide consulting and engineering services, since the Chinese system was so poorly engineered that it carried less than a third of its potential capacity.

Why wouldn't the Communist government want advice and counsel from the people who had long managed the biggest and best telephone system in the world? How could a businessman keep from dreaming about that opportunity?

When Klitgaard flew home at the end of the week, Sledge went to Hong Kong to meet with consultants who might be able to help Pacific penetrate the Chinese bureaucracy. There his staff reached him with an urgent message. Ms. Lee had information that a large Japanese telephone equipment manufacturer was on the verge of signing a major deal with China. Pacific would have to act quickly if it wanted to get into the game. Pressing her case, she wanted to meet Sledge before he left Hong Kong.[13]

Sledge was a risk-taker, but he wasn't at all certain this was a risk he should take. There were too many loose ends to the proposition. But the next day he met with Ms. Lee. She offered to help Pacific get its foot in the door, but now she wanted Pacific to send her husband to Beijing to work with Chinese technicians to make sure that Pacific's electromechanical switches would be compatible with the Chinese network. She also wanted the exclusive right to represent Pacific in China.[14] Still uncertain about her credentials and with no experience in this kind of international deal making, Sledge stalled. He asked her to provide him with a detailed explanation of the decisionmaking process in China. He wanted to know where the pitfalls were and how Pacific should proceed. The situation still seemed bizarre. He represented what would soon be a Fortune 50 corporation offering enormous telecommunications expertise to the most populous country in the world. Who was his go-between, his broker? He told Ms. Lee that he couldn't act at once and would think about her requests on the flight home to California.[15]

Back in San Francisco, suffering from jet lag, Sledge spent two more hours on the phone with Lee the next day. She pressed him for a decision. This time she wanted expense money. Again Sledge hesitated, thinking of the comic strip hero Charlie Brown. He had the opportunity to be the hero or the goat, to make Pacific Telesis look brilliant right out of the gate of divestiture or to look extremely foolish and naïve. He wasn't working from the Bell System manual, but some of the Bell System's conservatism served him well in this case. Cautiously, he told Ms. Lee that Pacific Telesis might be willing to provide some money if she could deliver a more tangible lead and specific evidence of her ability to get Pacific in the door. She told him she would call him back the following day.

But she didn't call the next day. Or the next. A week and a half passed, and then she telephoned her contact on Sledge's staff. She was close to a deal, she said. She wanted an engineer from Pacific who understood the technology of the switching systems to fly to Beijing immediately. When he arrived, she would brief him on the deal and get his input. Now Sledge was even more skeptical, but his staff believed the opportunity looked good. Reluctantly, Sledge agreed. He sent two members of his team to the Chinese consulate in San Francisco to get visas.[16]

Even as the engineers packed their bags for Beijing, however, Sledge and his staff were preparing their own China initiative. Pacific Telephone had gotten itself invited to a small telecommunications trade show in Guangzhou (Canton) in early September. It was "a two-bit" trade show, "a rinky-dink little catch-as-catch-can kind of thing," Sledge admitted. But it offered the company a chance to open another front, and there would be no other Bell System representatives at the conference.[17] Pacific would have a chance to plant its flag first in the mega-market of the next century. Sledge commissioned a video in Cantonese profiling the strengths of Pacific's technology and organization. He also ordered hundreds of color photos of the Golden Gate Bridge to give away.[18]

To Sledge's surprise, Ginn announced that he wanted to go along on the trip.[19] If Sledge had known more about his boss's childhood, he might have anticipated Ginn's interest in making this long and uncomfortable journey. Growing up in Alabama, Ginn had been fascinated by different cultures. He enjoyed his father's adventurous, low-cost family vacations to Mexico and Canada, where the people – as well as the sights – were different from anything he had encountered in Anniston. Since joining the Bell System, he had seen most of the United States, but he and Ann had done little traveling abroad. Once, in the mid-1960s, they took a trip with Ann's sister and her husband to Spain, where Ginn was amazed to see old men leading mules and carts, as well as the little trucks that serviced the government's poorly operating telephone system. On the Chamber of Commerce trip to China, Ginn had visited the Great Wall, the mountains of Kweilin, and the city of Beijing. That first seductive encounter with the potential of the Chinese market started Ginn dreaming.[20]

With his boss going on the trip, Sledge worked to get a top-level audience, one with the Beijing minister in charge of the telecommunications system. But negotiations through the American consulate stalled. Meanwhile, the woman from Pasadena became aware of these contacts. A week before Ginn and Sledge were scheduled to leave San Francisco, she sent word back to Sledge that he was "foolish" to pursue these plans without her. He was, she said, endangering her efforts to make a deal with the Chinese through the back door. For a moment Sledge was worried. But as he

Sam Ginn in China – At divestiture, Baby Bell executives like Sam Ginn dreamed of global operations, but PacTel's first marketing efforts in China in 1983 were hopelessly naïve.

told his staff, he was not prepared to turn Pacific over to the Pasadena connection. Unescorted, barely briefed, but determined, he and Ginn were going to China "to establish Pacific Telephone as a force" in the international arena.[21]

Ms. Lee was furious. She told Pacific's representatives that Sledge had betrayed her. She and Pacific Telephone quickly parted company, and Sledge heard the back door slam. He gave up any hope for a quick, earth-shattering deal. He and Ginn had learned their first major lesson in international business: though it was important to pay attention to events taking place behind the scenes, they now understood that it was foolish to think they could or should become masters of business intrigue. They would always have to knock at the front door. In China and elsewhere, he and Ginn would have to go "cold calling" to work their way into the market.

Before they left San Francisco, they arranged to be coached on the "dos and don'ts" of Chinese business etiquette. Enter a meeting with the most senior person first so that a staff assistant would not be taken for the CEO. Avoid raising new agenda items at a meeting, because "the Chinese don't like surprises." In negotiation, understand that few Asians will offer a direct "no" because they will not want their guest to lose face. In social situations, Americans accustomed to the clean plate club in their mother's dining room had to learn to leave food as a sign of politeness. They were coached on how to pace themselves when drinking *maotai,* a 120-proof liquor made from sorghum. They were also counseled on gift giving, a practice which was frowned on officially by both the United States and Chinese governments but which flourished in China and other Asian cultures and was deeply ingrained in the business and political systems. Eager to apply their new knowledge and confident that their energy and resources would make up for their lack of experience, the PacTel team left for China in the last week of August.

Their confidence quickly sagged, however, when they arrived in Guangzhou. Checking into the Deng Feng Hotel, an old structure built by the Russians across the street from the Exposition Center, they discovered soft beds covered with heavy, scratchy wool blankets. Rats skittered inside the walls. They had been warned that listening devices were probably installed on every floor in the hotel to record their comings and goings. Fearful of the water, Ginn and Sledge drank bottled water or beer. Both men wondered how they would ever be able to negotiate a contract in this setting – if, in fact, they ever got close to one.[22]

But on the first day of the exposition, their spirits rose. From the moment the doors opened, crowds descended on the Pacific Telephone booth. The stacks of photographs of the Golden Gate Bridge disappeared in minutes, along with the literature and ballpoint pens. "We had three days to go," Ginn said, "and nothing left to give away."[23] Over the next few days, he and Sledge met with dozens of officials in the Chinese telecommunications bureaucracy, pitching their "we can do for you what we have done for America" line. The Chinese asked lots of questions about the capabilities and engineering of the Bell System. Both Ginn and Sledge got excited.[24]

As the week wore on and there was still no news from Beijing about a meeting with the Minister of Post and Telecommunications, negative thoughts kept surfacing. Ginn began to get anxious about work back home. Divestiture was only three months away and there were a million last-minute disputes to resolve. He finally decided to leave early. But Ginn had no sooner broken Chinese airspace than Sledge heard from the American consulate that the meeting in Beijing had at last been arranged. Sledge called Ginn at his hotel in Hong Kong. "You have to come back," he said. Ginn hesitated and then agreed. The dream was too good to abandon so quickly. He spent the next several hours on the phone and at the Chinese consulate trying to get a new visa. To no avail. Finally, he called Sledge back. "Don," he said, paraphrasing Pogo, "I have met the bureaucracy and they have won."[25]

Sledge and one of the members of his team, a Chinese-American district switching manager, bought the only plane tickets they could get to

Beijing: one-way tickets with no guaranteed return. Their gamble paid off. The meeting with the Minister of Post and Telecommunications went well. They got down to the details that real telephone men enjoyed, talking about fiber optics, switching equipment, and network management. They were encouraged, even though getting out of Beijing provided another adventure. At the China Airlines ticket office they were told there were no seats available. They squirmed and finagled, using their slim list of contacts. Finally, associates at Jardine Matheson, the Noble House, got them seats usually reserved for Communist Party officials. Relieved to be on the plane, Sledge opened a copy of the English-language *China Daily* to learn that the Russians had blown up Korean Air Lines flight 007 the day before. News of the attack unsettled both of them. This definitely was not Kansas, or California.[26]

Despite the obvious risks of doing business in China, Ginn and Sledge were convinced by their experiences at the Guangzhou Telexpo that the tremendous Chinese market could not be ignored. Sledge wrote a ten-page business case promoting a new PacTel company to be called PacTel International (PTI) with a focus on selling telecommunications systems and services abroad. The business case targeted China and the Philippines for equipment sales. It also articulated a role for PTI as a consultant to national telephone companies. Positioning itself as an experienced operator steeped in the Bell tradition, PTI would teach telecommunications managers in developing nations how to run sophisticated networks and maintenance support systems. It could also provide purchasing advice and assistance with network design. Sledge thought this new international enterprise might be able to find opportunities abroad for PacTel's other diversified businesses in publishing and cellular.[27]

They presented the business case for PTI to Pacific Telesis top executives in November 1983. Don Guinn loved the idea. On the eve of divestiture, he said, it would fuel investors' interest in the company's stock. Perceived as the weakest of the Baby Bells because of its debt problems and vulnerability to California regulators, Pacific Telesis needed something to distinguish it from its six stronger sisters. John Hulse, the CFO, and Robert Dalenberg, the company's general counsel, were skeptical. The crusty Hulse said international looked like a hole in the ground that you dumped money into. He could not see a payoff within five years.

Dalenberg's concerns were even more serious. The MFJ precluded all the Baby Bells from entering any lines of business other than local exchange telecommunications without a waiver from Judge Greene. This rule gave Greene enormous power: it effectively made him the telecommunications czar of the United States. Dalenberg had been talking to the Department of Justice about a waiver for PacTel International since the woman from Pasadena first appeared on the scene four months ago. But the DOJ had made it clear that Greene did not intend to grant any waivers until the official date of divestiture – January 1, 1984. Even then, neither he nor the DOJ would look kindly on a sudden flood of applications. Dalenberg counseled caution. Pacific did not want to anger the judge, he said, by making a deal to sell equipment in China before the company had his permission. The whole diversification program might be jeopardized.

Ginn tried to counter Hulse's skepticism and reassure Dalenberg. He talked about what he and Sledge had seen in China. He was selling the China dream and he spoke with the enthusiasm of a true believer. He promised that no deal would be made before the judge had acted. By the end of the day, he had overcome Hulse's and Dalenberg's reservations, and his colleagues approved the creation of PacTel International.[28]

Sledge told Ginn that he should recruit someone with international experience to run the new business, but Ginn rejected that proposal. He wanted someone he knew and trusted. He didn't want a high-priced, international hot-shot – at least not at first, while they were still trying to feel out the potential for the business. "You're it," Ginn said to Sledge, "so get to work."

There was plenty to do. Even as they were pushing through the business case for the new international enterprise, clouds had gathered over their Asian initiative. Late in November, Klitgaard went back to China and met with a senior engineer from the postal and telecommunications administration. They discussed PacTel's proposal to provide consulting services to help China develop an integrated telecommunications infrastructure. They talked about PacTel's offer to sell "reapplied" No. 5 crossbar switching equipment. Klitgaard reminded the Chinese official that PacTel's switches would cost 75 percent less than the new digital switches being offered by NEC and Ericsson.

But as Klitgaard learned, price and pride were in conflict, and pride won. The Chinese engineer was unmoved. A 38-year veteran of the telecommunications bureaucracy, he had joined the agency shortly after the defeat of the Japanese in 1945 and remained through the Communist takeover of the government in 1949. He made it clear that his agency had little or no interest in buying Pacific's used equipment. The switches were not compatible with the Chinese system, he said. Connections would have to be retooled and personnel retrained if Pacific's switches were installed. The machines would be expensive to ship. Most importantly, China was not interested in a gradual improvement of its telephone system based on used American equipment. The government wanted to go directly to digital electronic switching, and it was looking for a technology transfer partnership that would allow China to begin manufacturing its own equipment. But because of the Bell System's historical dependence on Western Electric, Pacific had no experience in manufacturing, and besides, the antitrust settlement made it very clear that the Baby Bells were not supposed to get into the production of equipment.[29]

Despite losing that sale, Sledge and Ginn tried to keep the door open. They looked for any contract, no matter how small, that would give them a toehold. In December, as a follow-up to the Canton Telexpo, Pacific hosted a seminar for nearly 800 people from the Yunnan Science and Technology Exchange. The subject was crossbar switching systems. Traveling to Kunming – in the arid, mountainous region east of Burma and north of Laos and Vietnam, at a time when tensions between Vietnam and China had escalated into hostilities – Sledge could feel the percussion of mortar rounds exploding only a few miles away. Again he was uncomfortably aware of the enormous psychic and physical distance between this city in the interior of China and his home in the San Francisco Bay area.

Despite the ominous setting, the seminar went well. Pacific hosted a banquet for local Chinese officials that included Yunnan specialties: grasshoppers, frogs, small eels, and smelt. Out of this event, Pacific secured its first international contract, a $25,000 agreement to provide training in the United States for Chinese engineers. Eager to close his first deal, Sledge signed the contract weeks before the official date of divestiture and before Pacific had even filed for, much less received, a waiver from Judge Greene. Like most entrepreneurs, Sledge considered that legal process a mere technicality, something that could be tidied up later. But back home, the PacTel legal department exploded and dressed him down for drafting and signing the agreement without letting them vet it first. The clash between the entrepreneurial culture of the diversified businesses and the caution that had long characterized the old Bell System was becoming more intense.[30]

Neither the company's lawyers nor Sledge's precipitous nature were, however, Ginn's major worry in the first few months after the breakup – rather, Judge Greene was. As Ginn quickly discovered, the new czar of U.S. telephony was no more tolerant of the entrepreneurial spirit than was the Communist Chinese bureaucracy. Pacific filed its waiver application for PacTel International with Judge Greene in February. The application was one among nine the Baby Bells submitted within months of divestiture. Irritated by this rush to break out of a settlement on which the ink was barely dry, Greene reacted angrily. He accused the Baby Bells of seeking "to transform themselves from custodians of the nation's local telephone service into conglomerates for which such service [would be] at best a pedestrian sideline."[31] Greene's hostility set the stage for a fight between the Baby Bells and the company that was now their biggest competitor, supplier, and customer: AT&T.

The DOJ and AT&T echoed Greene's sentiments. The whole point of the breakup, they said, had been to keep the local exchange companies out of competitive businesses and thus to end forever the controversies over cross-subsidies and other anticompetitive activities.[32] The Baby Bells countered by asserting that Greene himself had recognized the need to allow the local exchange companies into new and dynamic lines of business. The activities they were seeking waivers for, they said, could not possibly be susceptible to monopolistic abuses. Robert Dalenberg suggested to Greene that the entrepreneurial spirit of the new companies should reassure him. These new ideas, he said, "are the fuel of the competitive process."[33]

Dalenberg's paean to capitalism did not arouse any cheers from the bench. Despite Dalenberg's sense that the oral arguments went well, he and the Baby Bells were stunned by Judge Greene's July ruling. Although Greene accepted the general tenor of the Baby Bells' arguments, he maintained, yet again, that the overall purpose of the settlement was to focus the attention of the Bells on local exchange telephone service. To ensure that they would stick to their knitting, Greene decided to set strict limits on their diversification efforts. No company, he ruled, could earn more than 10 percent of its revenues from diversified businesses. The judge also authorized the DOJ to review and inspect all of the new companies to make sure they weren't engaged in anticompetitive activities. Then, to Ginn and Sledge's dismay, he ordered the Baby Bells to rewrite and resubmit their waiver applications in line with these new guidelines.[34]

To Ginn, Greene's ruling made no sense. It seemed to turn capitalism on its head, restricting success while offering the Baby Bells unlimited license to fail. Judge Greene was insisting on slow, deliberate change in an industry experiencing an explosive transformation. By the time he finally granted Pacific its waiver for PacTel International in December 1984, Sledge and his team had been forced to tread water for almost a year. By that time, even Ginn was beginning to concede that the China strategy might never be a winner. Things moved too slowly on the mainland. Chinese bureaucrats were driven by the needs of status politics, not the kind of entrepreneurship that Ginn had studied in the Sloan Program at Stanford. "They would regard it as an admission of failure to contract with us," he said, "because it would indicate to the political establishment that they didn't know what they were doing."[35]

This was Ginn's second important lesson as he struggled to transform himself from a Bell-head to a global entrepreneur. Given that most of the phone systems in the world were operated by government agencies or public monopolies in the early 1980s, Ginn now understood that he would need an intimate knowledge of the bureaucratic players and local politics to build businesses abroad. For that, he and his diversified enterprises would need to obtain a strong local presence.

After expanding the original ten-page business case into a full-blown business plan for PacTel International, Ginn and Sledge set out to open offices in Malaysia, Spain, and South Korea before the end of 1984. They projected five more offices in Australia, Thailand, the People's Republic of China, Italy, and the United Kingdom in 1985. The goal was to start small, gain a foothold with a contract that got the company into the country – something like the training contract in Kunming – and then look for a major opportunity.[36]

Having solidified the support they needed in San Francisco, Ginn and Sledge for the first time began to reach outside the Bell organization to find key managers with international experience to run their offices. Working with a headhunter, they picked country managers for Korea, Japan, and Thailand who knew the players and the politics.[37] With these leaders in place, Sledge raced throughout 1985 to open offices, forge partnerships, and win foothold contracts. One of his first victories came in Spain, where PacTel and Telefonica, the national telephone company of Spain, agreed to develop a joint plan for a new telecommunications research facility in Madrid.[38] In Japan, PacTel opened an office and concluded a three-year agreement with Nippon Telegraph and Telephone Corporation (NTT), the national telephone company, to create an exchange program for personnel, information, and publications. Two months later, PacTel International began exploring business opportunities in India in collaboration with two other companies. These were important early victories.

Despite their lack of experience in international business, they were able to capitalize on their technical knowledge and the global reputation of the Bell System. It wasn't all bad news to be identified as a Bell type, especially overseas. As planning for the 1988 Olympics in Seoul, South Korea, got underway, PacTel recognized another opportunity to leverage its experience in Los Angeles. The company landed a preliminary consulting contract with the government of South Korea in November 1985. Meanwhile, PacTel had moved into the United Kingdom and become part of what

Sledge called "the deal flow" by buying a small voice-messaging company called One 2 One. Although none of the deals were big moneymakers, PacTel International now had a sense of momentum that was satisfying to Ginn and Sledge.

TEN

Crisis at Home

While he was trying to promote initiatives abroad, Sam Ginn suddenly found himself mired in a regulatory crisis in California. As all of the Baby Bells launched diversification efforts, state and federal regulators became nervous. They expressed mounting fears that the companies would subsidize their new ventures with revenues and resources provided by ratepayers. State regulators were locked into a rate-of-return system and mentality even more tightly than Judge Greene. The FCC had tried to address these fears, first by compelling the companies to create fully separate subsidiaries to sell telecommunications equipment and second by forcing them to tell customers that they could choose among vendors.[1] Despite these safeguards, the California commission – especially its Division of Ratepayer Advocates (DRA) – became increasingly hostile to the diversification program of Pacific Telesis.

The California legislature had created the DRA (originally named the Public Staff Division) in 1984 to institutionalize consumer advocacy within the CPUC. The new division was also a product of organizational changes taking place in the agency that intensified the adversarial relationship between at least one portion of the CPUC staff and the regulated utilities. Historically, engineers, accountants, and lawyers had staffed the commission. The staff had been organized by industries: telephones, electricity, transportation. But commissioners saw a greater role for economists and financial managers as it became apparent that the traditional foundations for rate-of-return regulation were eroding and that regulators would increasingly be engaged in the process of structuring markets. This transition at the CPUC bred administrative and cultural tensions. As one staffer described it, the organization divided: "it was the engineers versus the economists."[2]

The Public Staff became the strongest voice for the tradition of consumer protection that had long been the hallmark of the CPUC. They resisted the Pacific Telesis diversification program out of concern that the company would misuse ratepayer funds. In particular, they argued that all of Pacific Bell's assets – including its personnel – were ratepayer assets and that ratepayers deserved to be compensated for their use by Ginn's PacTel companies. Compensation, they maintained, should not be just at a market rate. The staff wanted instead to use an arcane formula that sought to tabulate the ratepayers' historic investment in those assets. These issues flared over PacTel International.

Seeking to prove that PacTel was misusing ratepayer assets, regulators ordered copies of PacTel International's telephone bills and pored over the numbers looking for calls to employees at Pacific Bell. They then went through the time sheets for those Bell employees to see if that time had been billed back to PTI. In effect, the state now sought to turn the whole idea of rate-of-return regulation on its head. Historically, shareholders had been compensated by ratepayers for the use of company assets – trucks, cable, switches, and even personnel. Now the state argued that the ratepayer had a residual interest in the assets and should be compensated like an owner if the company found other uses for the assets.

Understandably, Pacific Telesis rejected this logic. The company argued that the assets of the corporation belonged to the shareholders. The diversified businesses were leveraging those assets in highly competitive environments. Their executives had to make decisions quickly and they needed to make sure they didn't tip their hand to competitors. If the CPUC began to extend its long hand into all of the transactions of the equipment, publishing, and cellular businesses, as well as any other new companies PacTel might launch, then it would slow the pace of decisionmaking to an agonizing creep. Business strategies and secrets would inevitably be leaked. Most importantly, regulators who took the attitude that the diversified companies belonged to the ratepayers and not the shareholders would be perfectly justified in appropriating profits from these businesses to subsidize local exchange service for residential customers.

Ginn knew this issue and its history all too well. For years regulators had taken profits from AT&T's Yellow Pages and long-distance businesses to hold down rates for basic service. These subsidies created enormous confusion and controversy over the true costs and pricing of telecommunications services, and they resulted in lengthy bureaucratic and regulatory battles. Ginn had taken part in those battles during his tenure in Washington, DC, with AT&T. He had no desire to see the diversified companies forced onto this ancient battlefield.[3]

Flush with the freedom that divestiture had promised, Ginn decided to make a stand. When CPUC auditors came to the office to look at the diversified companies' strategic documents, his staff turned them away.[4] That hard-line stance, of course, infuriated the CPUC's agents. One commission attorney called PacTel's position "preposterous." Ginn's action also brought pressure from colleagues in his own organization, as Telesis lawyers and Pacific Bell executives began to warn him that his belligerence could endanger Pacific Bell's request for a rate hike and might bring down a huge fine from the CPUC. Still, Ginn refused to retreat. As he increasingly realized, the delicate art of managing regulatory politics in a changing environment demanded calculated confrontations if he and PacTel were going to win enough degrees of entrepreneurial freedom to succeed in a competitive market.

This time, however, it began to appear that Ginn had gambled too much. The Commission's investigation dragged on through 1985 and cast a shadow over the entire diversification program. The Commission's staff issued a preliminary report in April castigating PacTel for its lack of cooperation, sloppy accounting, and misuse of "ratepayer assets."[5] The report led to further investigations. The California Legislature passed a new law strengthening the Commission's ability to audit the books of utility company affiliates. Inside Pacific Telesis, Ginn was sharply criticized by employees and several of his peers for his handling of the whole affair, especially after the CPUC in January 1986 penalized Pacific Bell $4 million a year for PacTel's lack of cooperation.[6]

As the Commission continued its pursuit of the PacTel companies, it developed harsh rules that made it virtually impossible for Ginn to hire present, or even newly retired, Pacific Bell employees. Opportunities for synergies between Pacific Bell's business and the diversified companies, including wireless, were evaporating. Together, the CPUC and Judge Greene appeared to be making untenable the assumptions built into the business plans of the diversified companies, forcing Ginn to reevaluate PacTel's basic strategy.

His concern for the diversified businesses, including PacTel International, was heightened by a commitment he had made to Don Guinn and the Board of Directors in 1983 that the diversification effort would make a bottom-line difference in Pacific's earnings within five years.[7] By the beginning of 1987, the three-year deadlines Ginn had imposed on the diversified organizations were also fast approaching. Most of his enterprises were still struggling, despite his efforts to work with their presidents to improve their operations and, in particular, strengthen their financial procedures. One by one, Ginn had begun to pull the plug on these executives as their three-year deadlines passed. He scaled back or shut down struggling operations in publishing, leasing, and real estate. He got out of the equipment business altogether. For Ginn and Pacific Telesis's shareholders, this refocusing of the PacTel companies involved writing off losses that eventually totaled nearly half a billion dollars. These were tough decisions.[8]

In the case of PacTel International, the combined pressures of the CPUC and John Hulse forced Ginn to put a lid on the company's growth. The China dreams had not materialized. As Ginn had written to Don Guinn in the fall of 1985, the Chinese bureaucracy was 2,000 years old and not about to change any time soon. It was prudent to keep a minimal presence in China, given the size of the potential opportunity. But as Ginn put it, "My personal belief is that if there is a payoff at all, it is long term."[9]

Months later, Ginn decided it was time for a fundamental reevaluation of PacTel International's strategy. He began to make cutbacks. Sledge resisted the slowdown, arguing that PTI would lose critical momentum. He was right about that, but Ginn was firm. When Sledge would not back down, Ginn decided it was time for him to go. He offered him a position elsewhere in the company, but Sledge had gotten his taste of the entrepreneurial life and he liked being a CEO. He opted to leave. Ginn replaced him and gave his successor a mandate to control costs and refocus operations. In the meantime, he gave serious consideration to a fundamentally new strategy suggested by an outside team of consultants.

To some, it appeared that PacTel's diversification effort had failed entirely, and PacTel was not alone among the Baby Bells in disappointing some investors. Almost all of the Bells had poured hundreds of millions of dollars into failed efforts to develop real estate, publish directories, or open retail computer stores. Analysts at the Gartner Group estimated that the seven Baby Bells collectively lost $565 million on their unregulated revenues of $2.3 billion in 1985.[10] In some ways, Ginn and PacTel were exemplary for the discipline they displayed in pulling the plug. In contrast, Nynex did not exit its money-losing retail computer store business until 1992.[11] All of these losses could be written off to normal entrepreneurial risk and to the price of "learning by doing," but neither investors nor regulators were comfortable with that.

They found it hard to understand that entrepreneurs need only hit one home run. At PacTel, Southwestern Bell, Bell Atlantic, and other Baby

Bells, the losses on failed businesses paled alongside the value created in the domestic wireless industry. PacTel's finance staff estimated the value of its wireless operations in 1987 at nearly $2.5 billion.[12] From CEO Don Guinn's point of view, the capital gains in wireless justified the entire diversification program. Wireless was a three-run homer.

Having won the game with that hit, Guinn prepared to retire. He had always said that he would do so at the age of 55, and he was a man who believed in planning. During his tenure as CEO, he had overseen Pacific's dramatic turnaround after the crisis of 1980. He had faced down AT&T during the process of divestiture and, after the breakup, had led Pacific Telesis through three years of remarkable growth. Between 1984 and 1987, Telesis profits rose from $829 million to $1.07 billion, and the stock split twice. By 1987, however, Guinn was nearing his 55th birthday and was certain that he had the right successor ready in the wings.

As early as 1982, Guinn had focused on Ginn as the strongest candidate to become CEO of Telesis. True to his Bell System heritage, he made Ginn's development as an executive a high priority. In choosing him to lead PacTel's diversification effort, Guinn reasoned that the experience would be invaluable to Ginn if he became CEO. He was right. Three years after divestiture, the lessons of entrepreneurial failure, regulatory confrontation, and wireless success had taught Ginn a great deal about the new landscape for competition in telecommunications.

Guinn made Ginn's succession imminent in the spring of 1987 when he promoted him to president and chief operating officer of Pacific Telesis. In the opinion of one analyst, it would be tough to fill Guinn's shoes because he had been both a superior strategist and a charismatic leader. Ginn told reporters he wouldn't even try to emulate his mentor. "We are different people with different styles. I hope mine would be, overall, as effective as his."[13]

On the eve of Guinn's departure and only a few months past Ginn's 50th birthday, Pacific Telesis faced two major problems. Pacific Bell, the firm's local exchange company and cash cow, was caught in the middle of a leadership scandal that produced front-page headlines and cost the company enormous amounts of goodwill. As CEO, Ginn would have to rebuild employee morale and public trust. Meanwhile, on the unregulated side of the company, Ginn had his own unfinished business. Telesis needed to refocus its diversification effort. If new leaders could not save the remaining subsidiaries, then the companies would have to be closed. In the meantime, the CPUC had made it clear that the diversified companies could not hope to grow by leveraging Pacific Bell's assets, including its employees and customer relationships. Instead, Ginn would have to build those businesses by exploiting the company's newly acquired strengths in wireless. That would not be easy to do.

The strategic choices he had to make were perplexing. Wireless pulled him toward entrepreneurship, confrontation with the regulators, and further diversification. But the Bell way of managing tugged him toward a conservative approach that would protect the needs of the local exchange company. Then, unexpectedly, a political leader he had never met opened the path to a bold new strategy.

ELEVEN

Margaret Thatcher's Revolution

Ann Ginn watched Secretary of State George Shultz dance with actress Loretta Young as the couple twirled around the White House ballroom to the tune of "Hello Dolly." Despite passing her 76th birthday, Young was still beautiful, and Shultz was amazingly graceful for a man of his size and age. The ballroom, with chandeliers ablaze, was packed with men and women in tuxedos and gowns for the last state dinner of the Reagan era. Over her roasted saddle of veal, Ann marveled that she and Sam were there to witness this grand scene.

Seven months earlier, Sam had become CEO of Pacific Telesis Group. In some ways, this dinner marked a watershed in the Ginns' lives as well as that of the nation. Friends had warned Ann that with Sam's promotion things would get busier than they had ever been, but she could hardly believe that was possible. Over the years, as her husband's responsibilities increased and as they trekked from city to city, she had struggled to help their children adjust to new schools and new friends. It had become harder as the years passed and the children grew older. In high school their sons, Matthew and Michael, had both gone through a rebellious phase. Their struggles helped persuade Sam to turn down Charlie Brown's request that he return to New York during divestiture. Since that decision, the boys had both graduated and gone on to college. When Sam became CEO, their daughter Samantha, the youngest, was in her senior year of high school. Given his new responsibilities and schedule, it made sense for the Ginns to move into San Francisco rather than continue living more than an hour away on the East Bay. But this time Ann dug in her heels. She insisted on waiting until Samantha graduated. Now Samantha, too, had left home to begin college in southern California.

Faced with an empty nest, Ann abandoned the suburbs for the city and joined Sam on at least part of his globe-trotting schedule. The dinner at the White House was just the beginning. They had been invited along with Pacific Telesis's long-time dean of government relations, Art Latno, and his wife Joan. For years, Latno had enjoyed a special relationship with the Reagans, a friendship that stretched back to Reagan's years as governor of California. As a result, Pacific Telesis had eight years of unusual access to the White House, access that had been important during divestiture. But this dinner was not about business or politics; it was a special night for old friends, or as one newspaper called it, a "sentimental send-off for the Reagan era."

Sam Ginn's eyes swept over the dinner guests. They included heads of state, baseball stars, cabinet secretaries, politicians, Supreme Court justices, entertainers, and even the evangelist Billy Graham. President Reagan had a sore throat that made it difficult for him to talk, but he managed to acknowledge in a croaking voice their presence and support. He offered special praise and thanks to his guest of honor, Britain's Prime Minister Margaret Thatcher. Thatcher responded to Reagan's remarks with warmth, recalling her joy at Reagan's election in 1980, "knowing that we thought so much alike, believed in so many of the same things, and convinced that together we could get our countries back on their feet ...."[1]

Ginn and many others viewed Reagan and Thatcher as political soul mates. Together they represented the triumph of a new conservative philosophy that, among other things, championed deregulation, the liberalization of trade, and the privatization of state-owned industries. Earlier that day, Ginn had met with FCC Chairman Dennis Patrick and the two had talked about deregulation in the United States.[2] In Europe, Thatcher frequently stood apart from her peers on the continent, many of whom still favored state control of basic industries. Thatcher was leaning away from European economic integration for the United Kingdom and demanding a more market-centered competitive economy than most continental leaders were willing to embrace.

As Ginn knew, the changes set in motion by Thatcher's reforms were already dramatically reshaping his industry in the United Kingdom and the European Community. The hardwired systems were slow to change. But in the United Kingdom and Germany, regulators were on the verge of opening the wireless industry to competition, and Ginn had decided to go after that business. He had no doubt that the other Baby Bells would be there, too, joined by AT&T and perhaps Craig McCaw. But as he and Ann savored their fairy-tale dinner at the White House, he had reason to believe that his recovering California Bell-heads might be able to hold their own in Europe.[3]

*

Like wireless technology, European privatization and deregulation offered a once-in-a-lifetime opportunity. The state-owned and pervasively regulated hardwired systems there had allowed a significant gap to develop between their technology and performance and that of the American telecoms. The size of the gap varied from country to country. But on balance, U.S. companies had a substantial advantage over European firms led by executives even less attuned to competition than the least resocialized, retrained Bell-head. At firms like PacTel, which was adjusting to competition by ardently resocializing, retraining, and in some cases replacing its own personnel, the European gap in hardwired telecom systems looked large and attractive. It was not at all clear that European entrepreneurs would be able to respond to this challenge quickly, and Asian firms were unlikely to be competitors. The way seemed open and Sam Ginn was ready to guide PacTel International into a European market that the Bell System had abandoned long before either Ronald Reagan or Margaret Thatcher had developed aspirations to lead a global conservative movement.

Reagan and Thatcher drove deregulation and privatization ahead, but the seeds of change in telecommunications had actually been planted before the 1980s. Critics of regulated "natural monopolies" and government-owned telecoms integrated with the postal systems began to voice their concerns in the United States and Europe during the 1960s and 1970s. National security issues no longer served as an adequate defense of public ownership.[4] In Europe as well as America, economic concerns dominated the resulting political dialogue.[5] The scholarly critique of government and its regulatory institutions gradually filtered through to the staffs of various agencies and legislators and then in a watered-down form to their bosses.[6]

Nobel Prize–winning economist Ronald Coase contended as early as 1960 that regulation was basically a problem of market failure. The cost of regulation represented the social cost required to fix problems in the market. These costs could be negotiated by economic actors in the market place or they could be distributed broadly through the community by government. Either way, as Coase pointed out, the social costs were the same. The key question for policymakers was whether the government or the market could manage these costs most efficiently. Coase's paper launched dozens of studies of regulated industries; in most cases, the conclusions about regulation were negative.[7]

As we have seen, this new orthodoxy and the pressure exerted by potential new entrants into telecommunications markets loosened and then eliminated AT&T's monopoly and had repercussions around the globe.[8] Governments in Europe and other parts of the world, under pressure from large users of telecommunications equipment and services in their own countries, began to look more favorably upon competition.[9] U.S. trade representatives pressed for private rather than public systems. The United States, after all, saw its trade balance in telecommunications equipment shift from a $1.1 billion surplus in 1978 to a $2.6 billion deficit by 1988, and the government was anxious to open new markets for American equipment manufacturers.[10]

In Britain, privatization stopped being an academic issue when Margaret Thatcher came to power in May 1979. She and her Conservative Party supporters had an open disdain for government ownership and regulation and a powerful belief in the value of competition in relatively open markets.[11] They set out to revive Britain's anemic economy by shrinking what they characterized as a bloated public sector. At the time, a Post Office bureaucracy operated the telephone system throughout the United Kingdom. Fiscal politics and bureaucratic infighting had long delayed the kinds of investments in modernization that Ginn had overseen in Los Angeles and the rest of California. As a result, the British network was woefully out of date. Customers experienced long waits for service, and in all of Britain the quality of that service remained poor.

Convinced that modernizing the telephone network would strengthen London's position as a world financial center, the Thatcher government gradually privatized the telephone system in the early 1980s by separating the local and international networks.[12] That left the country with two government-owned companies: British Telecommunications Plc (BT) operated the domestic network, and Cable & Wireless provided international service.[13] Then the government opened the market for the sale of telecommunications equipment and licensed Mercury, a consortium of companies (including Cable & Wireless, Barclays Bank, and British Petroleum) to compete with BT by building and operating a second domestic network.[14] The United Kingdom became the first country to privatize a government-owned telephone company when it sold 50.2 percent of the shares of British Telecom to the public in 1984.[15]

With the United States and the United Kingdom leading the way, other governments began slowly to follow down the path to competition in telecommunications. The German Monopoly Commission challenged the Bundespost's control of the telephone network in 1981 in a public report that suggested that competition would benefit consumers and the economy. Similar reports appeared in Italy, the Netherlands, and Belgium.[16] Pressure for liberalization was thus building within the European Commission, which issued a "green paper" on the subject in June 1987. The proposal called for monopoly control in network infrastructure and voice traffic combined with competition in other telecommunications markets.[17]

The drift toward competitive markets gradually became a global phenomenon. Various factors – including the political strength of the national bureaucracy, labor resistance, and the ideology of the party in power – affected the speed at which individual governments liberalized and privatized their telecommunications markets. By the mid-1980s, however, the liberalization momentum was powerful and economists, journalists, and policymakers were optimistically championing the idea that market mechanisms would soon remove the need for regulation entirely.[18] As in the United States, the wireless industry became the driving wedge in the European movement toward telecommunications deregulation and liberalization.

∗

In the mid-1980s, Ginn and his team at PacTel were watching these events on the international stage very carefully. They were convinced that Europe would provide enormous opportunities for PacTel, something along the lines of the Chinese market but without the Chinese bureaucracy. In the summer of 1986, shortly after putting PacTel International's expansion efforts on hold, Ginn asked Bain & Company, a management consulting group, to evaluate PTI's strategy. The consultants' report, delivered that fall, confirmed the global trend toward competition and advocated a strategy of investing in and operating telecommunications and data transmission systems in other countries. As the consultants pointed out, PacTel's competitive advantage was in the "management" of ongoing systems and not in their construction. It made sense for the company to seek opportunities to own, or at least partially own, systems that would generate long-term revenues.[19]

The second part of the consultants' study focused on PacTel's problems with the CPUC. Given the commission's concerns about affiliate transactions, Bain said, PTI needed to focus on leveraging skills from the new diversified businesses instead of the skills and assets of the operating company, Pacific Bell. This meant looking for new business opportunities abroad in two main arenas: wireless and information services. In those arenas, PTI had already landed several promising contracts. In Kuwait, PacTel had entered into a joint venture in 1986 with the Japanese equipment manufacturer NEC to install a cellular system.[20] In February, the company had submitted a proposal to build a system in India as well.[21] In South Korea, PacTel had agreed to develop a credit-card verification system, and in Southeast Asia it had entered into an agreement with the Communications Authority of Thailand to install and operate the first digital display paging system in Bangkok. This project – Thailand's first major privatization effort – promised to give PacTel an inside position on the business opportunities generated by liberalization. In the United Kingdom, PTI had launched a voice-messaging service through its newly acquired subsidiary, One 2 One; this development, like the other new contracts, held out the potential for ongoing service revenues rather than one-time project management fees.

The Bain consulting report to PacTel coincided with a major shift taking place in the wireless industry abroad. While the FCC was still pondering how it would license the first cellular systems, other countries, less concerned about competition, had moved ahead using monopolies. Ericsson had built the first cellular system in the world in 1977.[22] Two years later, Nippon Telegraph and Telephone launched the first commercial system in Japan.[23] In Scandinavia, the Nordic Mobile Telephone System inaugurated service in Norway, Sweden, Finland, and Denmark in 1981.[24] All of these organizations were serving customers before the FCC even issued the first U.S. commercial licenses in 1982. In wireless, other nations had shot ahead of the United States largely because of regulatory sluggishness at the FCC.

The cellular networks abroad shared common features. They were analog and generally not compatible with one another. They were built for national phone companies that, in almost every case, extrapolated demand from their experience with existing mobile radiotelephone service. In many cases, these organizations were stunned – as were Ginn and his PacTel colleagues – by plan-breaking growth that quickly strained their ability to keep up with demand. In Scandinavia, for example, two years after the launch, Nordic Mobile Telephone System had more than 40,000 subscribers, and incoming orders for service were threatening to overwhelm the network.

Demand for the first generation of cellular services in Europe was strongest where the forces of competition were allowed to work – even if only as a duopoly. The number of subscribers soared to 1 million in Britain before Germany had even 100,000 mobile customers. The two British carriers, Vodafone and Cellnet, with growth rates between 25 and 30 percent, quickly became the largest mobile operators in the world.[25] As policymakers throughout Europe soon realized, competition spurred penetration and brought in tens of thousands of customers. In wireless, there was no shortage of entrepreneurial talent in Europe, and British entrepreneurs were particularly aggressive and successful.

Competition for customers did not, of course, displace the need for cooperation in dealing with governments. The rollout of different analog systems across Europe meant that customers (except in Scandinavia) were unable to roam from country to country and could not benefit from economies of scale in the manufacture of equipment. Prices for car phones and service remained high. By the mid-1980s, the analog systems were approaching capacity, and telecommunications managers and regulators were beginning to discuss the need to adopt a common technology that could meet the tremendous demand for cellular without consuming more radio spectrum. One solution was to reduce the size of the cells by adding more transmission towers. But radio engineers were hoping to find a way to get more efficient use of the available spectrum by fitting more calls onto one channel simultaneously. They were also hoping to meet customer demand for smaller, lighter, and less expensive handsets with longer battery lives. That called for a shift to digital technology.[26]

At this point, Europe took a turn that would move it a giant step ahead of the United States. In one of the most far-sighted moves by any government entity in the wireless sector – and, for that matter, in the entire history of telecommunications – the Conference of European Posts and Telecommunications organized the Groupe Spécial Mobile (GSM) to standardize the next-generation European cellular system in the 900-MHz bandwidth.[27] Through the mid-1980s, the GSM group worked on the development of this single digital technology and, by 1986, national governments were beginning to consider how they should license the new service. As in the United States, wireless would become the proving ground for the concept of competition in telecommunications.

Sam Ginn, who continued to watch these developments closely, was licking his chops to get PacTel into the European market. Just as he was settling into his new job as CEO of Pacific Telesis Group, however, he once again found his attention drawn away from new entrepreneurial ventures and toward a crisis at home. This time it was a crisis that some said could only have happened in California.

TWELVE

Deconstructing Culture while
Reconstructing Leadership

Ginn and other telecommunications executives decided in the 1980s that the traditional Bell style of management and the Bell culture were becoming outmoded – left behind by a fast-changing business and technological environment that was, in the jargon of the day, "challenging." But at AT&T and all of the Baby Bells, the old values were deeply ingrained in the personnel, from the top echelons to the shop floor. That was certainly true at the Pacific Telesis Group and its subsidiary Pacific Bell. At those organizations, management decided that only radical measures could transform Bell-heads into the market- and consumer-oriented competitors needed to deal with this new environment. In other industries as well, executives facing intense global competition were trying to deconstruct and reconstruct their corporate cultures. Some were attempting to copy the successful efforts of Japanese corporations to build interdependent teams with strong group values. Across the country, American companies trotted employees out to parking lots to perform calisthenics or packed them up for survival treks in the wilderness to build esprit de corps. On the eve of divestiture, Pacific Bell had launched a similar experiment with a program called "Leadership Development."

At first, Pacific Bell's executives were impressed with the results. They believed that Leadership Development improved work habits, enhanced cross-functional communication, and improved relationships between labor unions and management. But then problems started to develop. Work groups that had been through the training tended to separate themselves from groups that had not. Some managers pushed their employees to complete the program quickly – often faster than the program developers had envisioned – in order to remain in the good graces of their supervisors.

Employees began to complain about the arcane language of the program. They felt that they were being coerced into adopting a new way of thinking. Eventually, they took these complaints to the newspapers.[1]

When the press broke the Leadership Development story it created a new crisis for Pacific Telesis. The *San Francisco Chronicle*'s front-page article in March 1987 portrayed the program as a dangerous New Age management initiative that bordered on mind control. As the *Chronicle* pointed out, this radical, undemocratic program was costing California ratepayers "upward of $30 million a year." Its "cultish overtones" offended many employees. The article sparked yet another CPUC investigation. The Commission's Public Staff Division quickly concluded that the program wasted time and cultivated an atmosphere of "fear, intimidation and mistrust" among employees.[2] In the wake of the scandal, the president of Pacific Bell retired. At that time, Ginn was the newly appointed chief operating officer (COO) of Pacific Telesis and the heir apparent to the CEO's job. He worked with Don Guinn to make a series of critical changes in leadership that he hoped would diffuse tensions around Leadership Development and look down the road toward the development of his own successor. His decisions stunned many long-time employees.

To most of the people at Telesis, the natural successors to run the PacTel Companies and Pacific Bell were already in place in 1987. As Ginn's number-two executive, Phil Quigley seemed like the heir apparent to Ginn's job at the PacTel Companies. Clearly he loved the dealmaking and the competitive environment. He had never seen himself as a regular telephone man. Ginn, however, saw things differently. Quigley's experience with competition, his lack of baggage at Pacific Bell, and his potential to succeed Ginn as CEO of Pacific Telesis were all factors suggesting that he become president of Pacific Bell. There was considerable risk in making this choice, given Quigley's lack of experience with local exchange operations, and this was a risk the Bell System probably would not have found acceptable. But Ginn was looking forward to an increasingly competitive era in every phase of telecommunications, an era that would no longer be dominated by engineers and engineering values.

On the other side of the house, Ginn prepared to make an equally unexpected move. Lee Cox looked like a strong candidate to lead Pacific Bell. Like Ginn and Quigley, he had come into the Bell System in the early 1960s. After receiving an undergraduate degree in public relations from San Jose State, he spent his first year in the System writing press releases and occasional speeches. Cox's boss recognized his potential and made it clear that, if Cox wanted to go anywhere in the Bell System, he had to move out of staff work and into operations. He maneuvered Cox into a transfer to the business office, where he spent much of his day fielding complaints and collecting bills. The experience was formative. "It taught me this kind of grass-roots appreciation for the work that the real workers of the company did."[3]

A charismatic individual with a gift for motivating people, Cox developed strong relationships with his mentors. He rose quickly through the ranks despite his disdain for Bell System routine. Like other Pacific Telephone employees, he resisted a move to AT&T. When the breakup was announced in 1982, Cox (unlike Ginn and most other Bell-heads) was pleased. "I loved the whole idea," he said. Becoming an officer in the company the following year, Cox helped lead Pacific Bell's efforts to reduce staff, increase efficiencies, and become more focused on marketing. He became a key supporter and champion of Leadership Development.[4] Given the uproar that followed the *Chronicle*'s investigation into Leadership Development in 1987, Cox would have been a controversial president for Pacific Bell. But neither Ginn nor Guinn wanted to waste his ability as a leader. So they made the switch. Phil Quigley left his position as president of the PacTel Companies and took over at Pacific Bell. Lee Cox, who had been the heir apparent at Pacific Bell, replaced Quigley as the new head of the PacTel Companies. Once more, Ginn took a chance. This time he took that chance with cellular, the firm's most promising entrepreneurial venture.

With pressure mounting from investors, Ginn told Cox that he had only two years to bring the PacTel Companies into the black. To do so, Cox and his team continued the expansion of PacTel Cellular in the United States, acquiring new licenses in Dallas–Fort Worth and focusing on the development of digital systems in southern California. They took PacTel out of the equipment business altogether and scaled back the company's ventures in leasing and real estate. For a while they considered a move into the cable television business overseas and even bought some interests in the United Kingdom. This new growth path reflected their belief – and that of many others in and around the industry – that telecommunications and cable television were converging. Ginn saw these acquisitions as a way to get a better grip on the dynamics of the industry and to carry the organization further into international markets. He reinforced their strategy by providing them with another talented and experienced executive.

✳

Through most of 1987, PacTel International had been treading water. The company reduced overhead costs for business development and focused on projects it already had going in Spain, Thailand, Korea, and Kuwait. The company dumped two of its struggling subsidiaries, including the voice-mail service in the United Kingdom. During this period Ginn was looking for an executive with deep international experience who understood how to build partnerships abroad and win government contracts and licenses. He found the person he needed in his own organization.

Jan Neels was a quintessential internationalist. Belgian-born and -raised, he grew up learning four different languages in school. The son of a stonemason, he had an intense personality and a tendency to obsess on problems. After he received a degree in engineering, he went to work in 1962 for International Telephone & Telegraph (ITT) designing circuitry for switching equipment. Over the next several years, he became more involved in sales, which meant dealing with government or publicly sanctioned monopoly telecommunications companies. He traveled outside Europe for the first time in 1968, spending six weeks in Thailand bidding for a contract with the government. By this time, he had established his ability to move into the executive ranks, and ITT took him out of engineering and into project management. The company sent him to Venezuela for four years, with his wife and two sons, to manage a large project. Successful there, Neels became director of marketing and sales for the entire Asia Pacific region of ITT. He and his family lived for two years in Hong Kong before moving to Indonesia. As a country manager in Indonesia, he was in essence a conglomerate executive, dealing with a wide range of ITT products and projects. He became skillful in managing work groups that included in-country nationals and expatriates.

Neels enjoyed living and working abroad, but by the time his sons were finishing high school, he wanted to be closer to them while they attended college – whether in Europe or the United States. ITT accommodated by offering him a position at the company's New York headquarters, where he was responsible for the marketing and sales of telecom equipment throughout the Asia Pacific region. The business was booming as electronic switching replaced the old electromechanical equipment. But as Neels could see, ITT's product development was not keeping pace with the competition. By late 1983, he decided that the company was going to get out of the business, and he began to look for another job. He left ITT and for two years worked for U.S. Telecom, the predecessor to Sprint, in Atlanta. He was still in Atlanta in 1985 when Don Sledge's headhunter contacted him.[5]

Sledge told him that he was looking for people who could go into a country "cold," open doors, establish relationships, and make deals – especially in Asia, where Neels had extensive experience. Attracted to the opportunity because it had an entrepreneurial thrust, Neels took the job. When he arrived in January 1986, PacTel International was still a fledgling enterprise, and in his first year Neels whole-heartedly embraced Sledge's strategy of selling project management services to foreign telephone bureaucracies. But in 1986, after PacTel won the Thailand paging license, he changed his mind.

Neels was tired of selling hardware and project management. There were too many uncertainties – cultural differences, labor costs, suppliers, personalities – and when the project was done you had to go out and start all over again. Network-based transmission service was different. "It's a gold mine," he thought, "once the system is up and running." As he watched the first moves toward privatization taking place in Europe and Asia, Neels saw opportunities to start mining gold around the world. He got a chance to pursue this new strategy in December 1987 when Ginn promoted him to the presidency of PacTel International. Neels promptly headed off for Europe.[6]

With Neels fishing for new opportunities in Europe, Cox moving PacTel into the black, and Quigley doing damage control at Pacific Bell, Sam Ginn had the leadership team he needed to launch Telesis and its subsidiaries into a new era of telecommunications expansion. Wireless led the way, with Jan Neels sifting for the partners who would enable PacTel International to penetrate Europe's most promising market.

THIRTEEN

A Digital Home Run

The German government had issued its first cellular license to the state-run Deutsche Telekom, but then, in a bow to the new age of competition, the government decided to prepare for the transition to GSM by issuing a second cellular license (D2). The opportunity was huge. West Germany had the third largest economy in the world in 1988, with a per-capita disposable income nearly 30 percent higher than other countries in the European Community. Germans were known to be early adopters of new technology, with high standards for engineering and performance. As Neels and his colleagues studied the situation, they concluded that Germany was likely to be one of the best cellular markets in the world.

Since D2 would, however, be the first competitive license issued on the continent, they knew that the application process would attract the best companies in the world.[1] After reviewing the contestants, Neels and his team concluded that they could never win the license on their own. That would be politically unpalatable to the government. PacTel, they decided, should probably not even be the majority shareholder in a partnership. Fortunately, their experiences in Thailand and Korea had accustomed them to complex forms of partnership, and they were ready to accept a junior role in a successful combine so long as certain conditions were met.[2] Neels sent one of his staffers, a former corporate strategist named Ron Boring, to Germany in order to get the lay of the land and to begin identifying potential partners. Boring spent the early part of 1988 contacting BMW, Bosch, Daimler-Benz, and other leading German industrial companies. Daimler-Benz and BMW were considered to have the inside track on the bidding. BMW's management was outstanding, and the company had long been a German powerhouse.

But as Neels explained to Ginn and other top executives at Pacific Telesis, BMW wasn't his first choice. Nor was Daimler-Benz. He reasoned that the German government would find it hard to choose one of these firms over the other. Moreover, both were based in the affluent southern regions of the country, while the government was looking for a way to support the beleaguered industrial economy in the north, particularly in Westphalia and the Ruhr Valley. Ginn was impressed with Neels's ability to get inside the heads of the decisionmakers. When Neels suggested partnering with the old-line steel company Mannesmann, headquartered on the Rhine River in Dusseldorf, Ginn agreed to give it a try.

Founded as a seamless steel tube manufacturer in the Ruhr Valley in 1890, Mannesmann had become one of Germany's outstanding industrial firms by the middle of the twentieth century.[3] Through two world wars its factories had pumped out raw steel and industrial tube and pipes to build and rebuild Germany's industrial infrastructure. In the 1960s, however, the firm's executives had become discouraged with steel's cyclical swings and had started diversifying into capital goods and automotive parts, acquiring subsidiaries in precision instruments, heavy construction, information processing, pneumatic and hydraulic controls, and automated materials handling. By the late 1980s, these industries were also in trouble, but Mannesmann was well positioned to once again adopt a new strategy.[4] With nearly $2.2 billion in the bank, CEO Werner Dieter set out to find new growth sectors.

Information is as important to entrepreneurship as it is to the newspaper business, and Jan Neels just happened to have a good friend at Mannesmann who was working on diversification. Peter Mihatsch and Neels had worked together at ITT, and Mihatsch had a rich understanding of telecommunications. He had only recently joined Mannesmann, and in his job interview he had suggested to Werner Dieter that the company should consider bidding for the German cellular license. Dieter knew nothing about cellular, but he was intrigued. He decided to pursue this new opportunity even though it could be risky and expensive.[5]

Neels and Mihatsch chatted about a possible partnership at a not-really-casual casual meeting in Antwerp early in the summer of 1988. Neels explained PacTel's interest in Germany, and Mihatsch, in turn, talked about Mannesmann's need to push forward with diversification. The two men agreed to organize a more substantial meeting later in the summer. Each would bring a team. Neels explained that PacTel was talking to a number of potential partners, and Mihatsch of course countered that Mannesmann was doing the same.[6]

All over Germany, courtship dances were taking place as potential bidders tried to line up their partners. Liberalization in wireless gave the dances an international flavor, with cross-border alliances and joint ventures replacing the national monopolies (the public telegraph and telephone organizations – the PTTs) that had long controlled European telecommunications.[7] PacTel orchestrated a series of meetings with possible German partners in the summer of 1988, including another meeting with Mannesmann. By this time PacTel had a short list of four or five companies. Everyone in the business development group rated Mannesmann number one. But Mannesmann was wavering. Neels thought Mihatsch favored PacTel as a partner, but Werner Dieter was apparently looking elsewhere, to England.[8]

The British firm Vodafone had attracted Dieter's attention. A fast-growing product of the Thatcher reforms, it was virtually the only non-PTT cellular company in Europe in 1989. Swedish cellular pioneer Jan Stenbeck, together with Gerald Whent at the British defense contractor Racal, had organized Vodafone in the summer of 1982 after Stenbeck proposed that they collaborate on an application for the U.K.'s second national cellular license.[9] Whent had never heard about cellular. But as he listened to Stenbeck he got interested. The two firms applied for the license and won. Whent then toured the United States, visiting cellular operations at Nynex, Ameritech, and Bell Atlantic before ending up in Los Angeles at PacTel, whose operation impressed him the most. When he returned to the United Kingdom, he told Racal executives, "I think this is going to fly." After launching service on New Year's Day in 1985, Vodafone set a pace that could not be matched by the slow-moving British Telecommunications. Within a year, the new firm had raced ahead of Cellnet, the cellular subsidiary of the former government monopolist.[10]

Vodafone had contacted Mannesmann. While Chris Gent of Vodafone had been impressed by Mihatsch's energy, he regarded Mannesmann as "a lumbering old, rust-belt company" full of engineers who didn't understand marketing. He was much more impressed with BMW, whose management ranks ran deep with talent. Moreover, in 1989, according to Gent, "Car phones were much more central to our thinking than hand portables, which were still big and clumsy." An alliance with one of the premier automobile companies in the world made sense.[11]

With Vodafone leaning toward BMW, Dieter listened again to Mihatsch, who was arguing strongly for an alliance with PacTel. Dieter finally deferred to Mihatsch's experience in telecommunications and opted for the U.S. alliance.[12] If they could agree on a joint venture with PacTel, it would be named Mannesmann Mobilfunk, and both partners knew when the negotiations started that PacTel would have to take a minority position in the business.

Still, PacTel wanted to be protected. Ginn and his colleagues contended that, if they were going to commit hundreds of millions of dollars to the construction of a new German system, then they had to have some measure of control. Initially, Mannesmann proposed to take a 74-percent share of the business and suggested that the enterprise should be governed by a simple vote of the shares. On certain critical decisions, PacTel countered, this was unacceptable. PacTel demanded veto power over the business plan and the selection of key personnel. Since PacTel would shoulder primary responsibility for the design, construction, and management of the network, Ginn, Lee Cox, and Jan Neels wanted the right to appoint the senior technology officer. Given the German company's lack of telecommunications experience, these conditions made good sense, but still the two sides haggled. A touch of national pride came into play, but eventually Mannesmann agreed to these terms.[13] In October, the two companies signed a formal memorandum pledging to work together on a joint application.[14] Ginn had good reason to be pleased.

Throughout the entire period of the corporate courtship and negotiations with Mannesmann, PacTel was also wooing the German government. In the United States and the United Kingdom, competition in telecommunications was broadly accepted by 1989. But in Germany there was still a great deal of hesitation and doubt about the new world of global partnerships and intense competition. This wasn't the type of political economy that had been a crucial element in Germany's success during the Second Industrial Revolution. The new style of business seemed, from a German perspective, to be messy and unpredictable. But it also seemed to be the wave of the future in Europe, where Germany wanted to continue as a leader. As the country's bureaucrats prepared the ground rules for competition, they wisely solicited input from existing cellular providers on engineering standards and evaluation criteria. These discussions were extremely helpful for PacTel, which hoped to emphasize throughout the process its technical strengths. The discussions helped focus the Mobilfunk team on critical issues as they worked seven days a week in Mannesmann's training center in the Black Forest to prepare the application.[15]

Their strategy for winning the D2 license reflected PacTel's experience in the cellular industry and its understanding of the mistakes that incumbents were prone to make. Mobilfunk intended to impress the German bureaucrats with their professional and technological competence. They planned to use all of the 1,000 pages allowed, including as much detail as possible. They outlined an aggressive build-out schedule with solid financing to support a worst-case cash-flow situation. The system would provide service to portable handsets right from the beginning. The partners agreed to overbuild the network to ensure sufficient capacity. This aspect of the plan was crucial because Deutsche Telekom was already having problems keeping up with demand.[16]

While the proposal team ground ahead at Mobilfunk's special command center, Neels and Mihatsch stewed over the politics of the application process. They decided to strengthen their hand by broadening their alliance. They brought Deutsche Genossenchaftsbank into the venture for 10 percent of the equity. They then covered another flank. Vodafone had joined the BMW team, which included the French water company Générale des Eaux SA (later known as Vivendi). Concerned about British political support for Vodafone's application, Neels and Mihatsch brought in the U.K. firm Cable & Wireless for 5 percent to neutralize the British trade representatives. Late in the process, they filled out their combine with a French company and a German craft union. This idea was Mihatsch's, who was tugging on what he called "the social element." Ginn was mostly worried about competition from PacTel's siblings, the other Baby Bells, all of which could claim to have similar technical capabilities. Nynex had allied with British Telecom and Daimler-Benz. BellSouth ended up with Siemens, Olivetti, and Shearson-Lehman-Hutton.[17]

At Mobilfunk, the Germans and the Californians worked well together, but Ginn and his colleagues still had a great deal to learn about doing business the European way. Following the American model, the PacTel team wanted to carry its own lobbying effort to Bonn. Mannesmann wisely dug in on this point and made it clear that if anyone was going to lobby the German government, it would be Germans. There were other points of authority that had to be established. For instance, as the application process came down to the wire, the technical specifications written by PacTel's engineers in California had to be translated into German. A team of twelve translators pored over the faxes as they came in from the United States. PacTel, nervous about the translation, insisted that the final document be retranslated into English for proofreading and fact checking. The Germans pointed out that it would be nearly impossible to find enough translators familiar with the arcane vocabulary of telecommunications to complete this work in such a short time. They invited the Americans (with tongue in cheek) to feel free to try to do it at home.[18]

As the engineering studies were being completed, the price tag for the infrastructure, including hardware and software, kept increasing. When it approached 4.5 billion DM (Deutsche marks), the finance department at Mannesmann panicked, warning Dieter that the company did not have the cash to provide its share of the equity. Still hesitant about the potential for cellular, they contemplated a palace rebellion. But during a meeting in Germany, Lee Cox provided an overview of the tremendous success of PacTel's cellular operations in the United States. He made it clear that wireless was "a cash cow" that was generating margins as high as 40 and 50 percent in some markets. Cox's presentation diffused the insurrection. The fears of Mannesmann's financial team were also eased when a consortium of German banks stepped forward to provide a 1.56-billion DM line of credit.[19] For the moment, Mobilfunk – Mannesmann's most expensive diversification effort – was moving forward.

Competition for the D2 license was fierce. Seven different consortia – including five other Baby Bells as well as Bell Canada Mobile, GTEC, Mc-Caw, and more than a dozen other companies – filed applications. Each ran nearly 1,000 pages and cost millions of dollars. For the companies entered in this German variation on a "beauty contest," the effort represented an enormous roll of the dice. Rumors were rampant through the fall of 1989 while the Mannesmann Mobilfunk team waited for a decision. The German government broke the suspense on December 7, 1989, with an announcement that what Craig McCaw later called "the best cellular license in the world" was awarded to Mannesmann Mobilfunk.

Neels's political calculations had apparently been right on the mark and, four days later, he and PacTel scored big in Europe again. The British government, staying at the forefront of competition in cellular, had decided in the summer of 1989 to award three licenses for a new wireless technology called Personal Communication Network (PCN). Most cellular customers were still tied to their cars because of the cost and weight of portable phones, but PCN promised to change that situation. Utilizing lower-power systems that could deploy lightweight, affordable phones with smaller batteries, PCN offered cellular-like service to customers in dense urban environments. Some people believed that these new networks might eventually steal significant market share from existing cellular providers and even from the wireline market if customers embraced the idea of having one number, in one phone, that would go anywhere.

PacTel had joined a team led by British Aerospace to bid for one of the PCN licenses, and in December 1989, Ginn received good news. Their joint venture had won. PacTel suddenly had the opportunity to get the jump on its U.S. competitors by launching two digital systems well before any of the domestic carriers made the conversion.

But the U.K. license also forced Sam Ginn to confront some harsh realities. With the inauguration of PCN, the United Kingdom would become the first major market to experience the kind of wide-open competition that would narrow margins and make mistakes costly. While Ginn's organization had acquired a great deal of capability in conducting international operations, PacTel would be hard pressed to keep both of these complex organizations working smoothly all of the time – especially if Ginn couldn't find the capital needed to keep both ventures fully financed. He suddenly realized that PacTel may have been too successful in the beauty contests.

As economist Joseph Schumpeter had long ago explained, the central entrepreneurial problem was finding the capital needed to fund innovation. Ginn had already experienced and survived that problem in Germany. But by the end of the 1980s, the entire global telecommunications industry was entering a dangerous era in which capital constraints would shape and reshape the process of innovation. The demand for capital – needed to lay new cable systems, transform hardwire networks, get wireless running, and provide storage and switching for the Internet – would be enormous. The need for money and the cost of this capital would force some of the best-established organizations in the industry to alter their strategies, adopt new corporate structures, and tailor the pace of innovation to fit the availability of capital on reasonable terms. PacTel and all of the other players in wireless would experience intense pressure from global capital markets.

As Ginn well knew, the German D2 cellular license was no toehold project. Once completed, it would be the largest integrated digital cellular system in the world. Run successfully, it could make PacTel a major player in the European business. But if there were major delays in getting the system up and running, or if the new GSM technology didn't do what it was supposed to do, then Mobilfunk and PacTel would quickly be in billion-dollar trouble. The uncertainties in 1989 were formidable. The GSM standards had yet to be adopted. Equipment manufacturers like Ericsson and Motorola had neither transmission equipment nor handsets in production. As he catalogued the potential problems, Ginn considered the fact that Mannesmann and PacTel still knew very little about one another, least of all whether they could make their partnership work on a day-to-day basis.

When he flew to Germany to attend the ceremonies and celebration after the official award of the license, Ginn got a taste for what lay ahead. The speeches by Schwartz-Schilling, Werner Dieter, and others were of course in German, and Ginn sat at his place smiling, not understanding a thing. When it came time for him to speak, he smiled again and asked the group's permission to speak in English, "otherwise we'll be here all night while I stumble through in German." The audience was gracious. But as he sat down, Ginn felt like a bridegroom who has married into another culture as well as another family. The wedding had been a tremendous success, but now he had to make the marriage work.

Mannesmann's executives brought to the alliance their expertise in German politics, land use, construction, and business culture, but they knew nothing about wireless or telecommunications. As Mannesmann's project manager, Erhart Meixner, protested to his boss when he was first assigned to the project, "The only thing I know about telecommunications is how to pick up a phone and dial." "That's okay," his boss said. "Our American partners know this business."[20] But PacTel's engineers, who indeed understood telecommunications, were all Californians; many of them had never been out of the United States except to take a cruise in the Caribbean or cross the border south of San Diego. Most did not speak German, and they had little understanding of European history or politics. Even as they celebrated winning the German license, Ginn, Lee Cox, Jan Neels, and the other executives in San Francisco were trying to decide who should go to Germany to make certain the operation ran successfully.

At a Christmas party that December, Ginn joked about the problem with a group of career Bell System executives. Various men from European immigrant stock offered their best phrases in German, Dutch, or whatever continental language they could muster. George Schmitt, who was in charge of Pacific Bell's regulatory operations in California, won this off-the-cuff competition because his parents had been German immigrants. As it turned out, he got more than a good laugh.

Schmitt had grown up in New York City, but he could stumble his way in German because his parents spoke the language at home. He was much quicker with the vocabulary he had mastered as a U.S. Marine, including some colorful combinations of four-letter words. In the Corps Schmitt had learned how to take and give orders and how to get a job done, no matter what the price. Normally noncombative, he could nevertheless scorch a man's personality when the job was at stake. After his time in the service, he had risen through the ranks of the Bell System, rotating like other executives through various disciplines, including engineering, real estate, and operations. By Christmas 1989, as vice-president for regulatory affairs, he was in one of the hottest seats in Ginn's house. But within a month, he was winging his way to Dusseldorf.[21]

To George Schmitt and the other Californians who arrived in Germany in January 1990, the challenge of guiding a large-scale project to completion in a foreign culture and language seemed overwhelming. When engineer William Keever arrived in Dusseldorf later that month, he spoke only a few words of German that he had learned from a Berlitz guide. In a hotel during his first week, he tried to order ice for a drink and received a bowl of ice cream instead. It made him wonder how he was going to order the parts needed to build a multi–billion-dollar telecommunications network spanning all of Germany.[22]

The first meeting between the Mannesmann and PacTel teams didn't make him feel any more optimistic. Some of the U.S. executives, overconfident of their engineering knowledge and intending to run the project from California, seemed to be practicing for a role in the stage version of *The Ugly American*. They were, they made clear, already focusing on their own development of a new digital standard called CDMA, and they were openly skeptical of GSM. When the week was over, the PacTel Cellular executives went home, leaving Schmitt and his recruits to engage in damage control. They knew they were in Germany for the duration and were dependent on the Germans' knowledge of how to get things done in that country. With a bit of practical diplomacy and a cooperative attitude, they were soon able to clear the air and to forge a bond with Mannesmann's veterans, particularly Erhart Meixner, the project manager Mannesmann had picked to ensure the system got built.

Unfortunately, just as the two teams got down to serious work, California's executives intervened again. Initial progress on the project was slow, and that encouraged the president of PacTel's domestic cellular operation to intrude, wrestling for control of the enterprise. Tensions mounted between the home team and Schmitt's displaced Californians. It took Meixner's intervention to convince the intruders from home that they would have to rely on German expertise. PacTel President Lee Cox settled the dispute in favor of the team in Dusseldorf. This was a critical turning point for Mobilfunk, and the decision reflected both Cox's and Ginn's belief that operational control should be decentralized to those on the ground. Schmitt never had to look back to California for approval as he and Mannesmann Mobilfunk raced to erect cell towers and turn on the D2 system.

*

With the decision to delegate, Schmitt's problems and opportunities were firmly rooted in Europe, which suddenly began to produce both in extraordinary fashion. Along the Berlin Wall in November 1989, the East German guards marched away from their posts and didn't return. Within hours, German citizens swarmed over the barrier and then attacked the Wall with sledgehammers. Within a year, Germany was reunited and the Cold War was over.

The changes in political economy that followed this historic moment stunned nearly everyone. In parallel with the West's gradual turn away from state-owned enterprise and towards privatization and deregulation, Eastern Europe suddenly embraced competition and private enterprise in hopes of raising living standards. Capitalism had raised per-capita incomes in the United States more than tenfold in the course of two centuries, and citizens in the communist countries had grown impatient waiting for centralized bureaucracies to deliver similar results. Although the fall of communism gave East Europeans tremendous personal freedom and opened new markets to Western companies, the transition from communism to democratic capitalism was slow, difficult, and painful. On occasion it was also comical.

For Mannesmann Mobilfunk, unification immediately raised the issue of whether or not the D2 license would be extended to include East Germany. The company's erstwhile competitors pushed the government to award a new license for the East. Meanwhile, the East Germans expressed skepticism about whether the technology would even work. When Schmitt set out to demonstrate that PacTel's technology could do the job, he found himself forced to convince the East Germans that California was not some elaborate Hollywood stage set erected to fool their government.

This bizarre episode began at a five-star restaurant outside of Dusseldorf. Mihatsch introduced Schmitt to four members of the East German parliament. After a six-hour dinner complete with champagne, wine, whiskey, more champagne, and a seven-course meal, Schmitt said, "they wanted to know – because we didn't have our network up in Germany yet – whether it was real." Schmitt invited them to visit California. None of the East Germans had ever been out of the country, and they jumped at the opportunity to see Hollywood and the West Coast's other attractions. PacTel flew them to San Francisco, where they had rooms at the elegant Saint Francis Hotel on Union Square. Schmitt took the officials on a tour of the city, including a stop at the Safeway in the Marina District. "They had never seen anything like this. In East Germany, you went into a store and you bought what the butcher had. They weren't used to seeing 200 chickens and 200 steaks. They thought it was all a setup." That night at dinner in the Carnelian Room at the top of the Bank of America building, the East Germans couldn't take their eyes off the view of the bay, Alcatraz

Island, and the Golden Gate and Bay Bridges. "We had a wonderful time," Schmitt said. But still, the East Germans were suspicious. California was too good to be true. It had to be an elaborate ruse.

The next morning, the group flew to Los Angeles in a corporate jet for a tour of PacTel Cellular's headquarters in Costa Mesa. Lufthansa had lost the luggage of the representative from Leipzig, and Schmitt kept offering to take him shopping for clothes. The East German replied, "No, everything's okay." But on the trip to Southern California, his colleagues kept giving him a hard time about wearing the same clothes and he finally relented. Schmitt took him to the opulent Fashion Island mall in Newport Beach.

In the air-conditioned men's department, Schmitt asked, "Okay. What do you need?" The East German, a tiny man, asked for underwear. But the sizes didn't translate, so he had to be measured, and then he had to make a selection. "They must have had twenty different kinds," Schmitt says, "boxers, pink, white, yellow, green. 'Which kind do you want?' I asked." The East German pointed to regular briefs. Then he wanted pajamas. "What kind of pajamas do you want," Schmitt asked, "long ones, short ones, with a fly, with buttons?" "He was looking at me like I was crazy." By the time they left, Schmitt's East German guest was lugging an entire suitcase full of clothes.

Back at the office, he told the other men, "There's no way on earth this can be a setup." He described the mall, the cars, the clothes, the air-conditioning, and the people. "By then, it was over," Schmitt remembers. Except for one last step.

In the car one of the men asked: "How do we know all these wireless systems you show us work?" Schmitt handed a phone to the leader of the group and said, "Dial your office."

"It won't work," the German responded.

"Yes, it will," Schmitt replied.

So the man punched in the codes for international and for East Germany and then the number. The other men watched and waited. After a moment, the man's secretary answered. That did it. From that moment the East Germans were convinced of the power of the technology. While there were still many hoops to jump through on the way to winning the East German license, Mannesmann Mobilfunk persisted and scored another major victory.

During the transition from national telecommunications monopolies to global competition for communications services, many long-time telephone company bureaucrats – like Sam Ginn and Don Sledge in China, Bill Keever and George Schmitt in Germany, and the East German officials in California – created more than one comical moment in their efforts to venture beyond their own cultural horizons. The amazing thing is how quickly many of them adapted to different cultures and the rapidly changing paradigm for telecommunications.

The East German cellular license added to the D2 system another 20 million potential customers and some additional headaches in construction. This was clearly an enormous coup, equaling almost two thirds of the number of POPs covered by PacTel's domestic cellular companies.[23] But with East Germany added to the construction project in February 1991, Mannesmann Mobilfunk's problems in securing enough equipment and cell sites increased. The team had to bully suppliers and adopt an unusually creative approach to cell sites, installing transmitters on historic structures and buildings, including old barns and castles, to get the coverage they needed. Desperate for handsets, Schmitt told attendees at an industry conference that GSM really stood for "God Send Mobiles," and he passed out buttons with Mannesmann Mobilfunk's new slogan.

These problems were gradually solved. In June 1992, they were finally ready to launch the system. Ginn flew to Germany, where he was scheduled to place a symbolic call to U.S. Secretary of Commerce Barbara Franklin in Washington, DC. Schmitt and Keever had planned to make the call from the office. Ginn said, "That's crazy. This is a mobile phone, I want to be somewhere outside." So they drove out along the Rhine. Ginn stepped out into a sheep pasture followed by a photographer. Afraid that the battery wouldn't last, Keever watched nervously. The first call to Franklin was unsuccessful because she was on another call. Her secretary asked Ginn to call back in ten minutes. When he did and had to hold, Keever began sweating. Finally, they were connected. The minute Ginn said, "Well, it's been nice talking with you" and handed the phone back to Keever, the battery warning alarm went off![24]

At first, the German D2 system operated a bit like Ginn's mobile phone – on the margins of power. But when the network began commercial operation on June 16, 1992, Mobilfunk was able to cover all of the nation's major highways and population centers. Ginn and his German partners quickly discovered how easy it was to compete with the entrenched PTT. Deutsche Telekom had underbuilt its original analog cellular network, and by 1989 the system had reached full capacity with 130,000 customers. The overload meant that customers were frequently disconnected.[25] When D2 became available, these disappointed customers flocked to the new system.

Within six months of switching D2 on, Mannesmann Mobilfunk had 100,000 customers; by April of the following year, more than 120,000.[26]

"D2 Takes Off" – Depicting itself as a young fighter in this highly successful ad, Mannesmann Mobilfunk (a PacTel joint venture) brought competition to German telecommunications with the launch of its digital cellular service in June 1992.

Despite some early technical problems, they were ahead of their projections and quickly exceeded Deutsche Telekom's market share.[27] Cellular penetration rates in Germany lagged the United States: only 1.5 percent of the German population were subscribing, compared to 4 percent in

America. But Mannesmann Mobilfunk was in a strong position and remained confident about the potential of the German market.[28]

*

Success in Germany gave PacTel a lead on the competition for GSM licenses in other parts of Europe. With Jan Neels heading the business development effort, a license application group that fully understood the technology, and a growing knowledge of the character and potential demand for cellular, the U.S. firm was an ideal partner for European companies interested in getting into wireless. There were plenty of offers. So long as cellular licenses were being awarded by beauty contests, with governments evaluating the technical and financial capabilities of the bidders, PacTel's experience in Germany gave the firm a formidable presence.

The German experience opened the eyes of the world to the potential for international cellular operations. The pace of worldwide liberalization was accelerating. Under pressure from the European Union and U.S. trade negotiators, Europe's state-owned telephone companies were making plans to privatize. There seemed to be opportunities everywhere and, by the early 1990s, PacTel was competing against the other Baby Bells and AT&T in almost every corner of the globe. In Budapest, U.S. West signed an agreement to build the first cellular system in Eastern Europe. It later built the Soviet Union's first cellular system in St. Petersburg and, with Bell Atlantic, launched cellular service in Czechoslovakia. Meanwhile, BellSouth took an early lead in South America, and Southwestern Bell (SBC) bought a major stake in Mexico's telephone company, Telmex. Bell Atlantic and Ameritech joined forces to buy New Zealand Telecom for $2.4 billion in 1990. Although AT&T focused initially on selling telecom equipment abroad, it increasingly eyed the service sector as well. Its CEO, Robert Allen, pledged that AT&T would generate 50 percent of its revenues from offshore sales by the year 2000.[29] Market opportunities for all of these companies increased after negotiators included telecommunications services in the GATT trade agreement signed in 1994.[30]

PacTel's team in Walnut Creek, California, which oversaw preparation of the license applications, was running full speed trying to keep up.[31] PacTel bid for, but lost, a license in Denmark in 1991. In Portugal, a PacTel consortium known as Telecel won. That same year the company labored over a license application in Greece. The company won paging licenses in Spain and Portugal as well. It also formed a partnership to bid for a cellular license in the Netherlands. With its involvement in Germany and Britain, its paging operations in Thailand and Korea, its stake in a long-distance network in Japan, and its cable television assets in the United Kingdom, PacTel had emerged as the most international of the seven Baby Bells.[32]

PacTel's early success abroad derived from a combination of factors. The first was an accurate concept of what was likely to happen in Europe. As European governments created opportunities for competitors to enter wireless markets, Ginn saw what Craig McCaw had seen in the United States: a duopoly structure that delivered fat margins on competition against bureaucratic competitors that were likely to be slow-footed. In the competition for licenses, PacTel made good use of its extensive experience in regulatory fights with the CPUC and AT&T, fights that had made its executives particularly sensitive to government relations. George Schmitt came to Germany straight from a regulatory job working with the CPUC, an experience that helped him deal with German bureaucrats. In the era of "beauty contests," political skills mattered a great deal. So did technological experience in engineering and running a high-quality network. Successful applicants also had to show that they could team with local partners and that their team could provide the greatest benefits to consumers and the state. PacTel's failures in China and elsewhere had taught the company how to accept a minority role in a partnership while retaining control of key decisions that might expose the company financially or operationally. These organizational skills, learned sometimes through failure and always through struggle, helped PacTel succeed abroad.

*

Ironically, all this success created a new crisis at home for Ginn. Expansion had to be funded, as did further modernization of the California network. Cellular was burning capital at a tremendous rate, and it was not evident when it would make a contribution to the bottom line. The California Public Utilities Commission was meanwhile contemplating a new regulatory structure. With the breakup of the Bell System, the political momentum had begun to move away from rate-of-return regulation in the United States. Commissioners and operating companies were interested in "incentive regulation" that encouraged utilities to maximize productivity in exchange for a higher rate of return.[33] In California, the CPUC had issued a report in May 1987 suggesting that prices for the most competitive portions of the local exchange business would have to be driven down.[34] Pacific Bell had responded by proposing a new regulatory framework that would provide a target rate of return and share anything over the target between shareholders and ratepayers.

When Ginn became CEO of Pacific Telesis, the CPUC was already holding hearings related to the new regulatory structure. The hearings made it quite clear that competition was the quid pro quo that would be demanded in exchange for allowing Pacific Bell a framework that would give it a higher return on equity, uncapped earnings, and a floating debt ratio. Given the productivity gains that Pacific Bell had made over the previous decade, Pacific Telesis had high hopes for this new structure, even with competition. The company accepted a major rate decrease that took over $300 million out of its annual revenue stream to win approval for the new setup. It also accepted productivity targets that were very high.[35]

Ginn and others believed they could hit those targets. Then, a nationwide recession took hold in 1990 that slammed the California economy late in the year. As the recession continued into the following year, Ginn realized that Pacific Bell's earnings would not improve any time soon. It didn't help that Pacific Telesis was not the only Baby Bell confronting these issues. A decade after Charlie Brown's announcement that AT&T would divest the local operating companies, the stock values of all the Baby Bells were depressed by the threat of deregulation and the prospect of competition in the local exchange.[36] This was not a new story. Investors and Baby Bell executives had worried about this scenario since the breakup. But as long-distance companies, cable operators, and upstart local access firms took aim at the local exchange, investors grew extremely nervous.[37] The need to make major investments in local exchange infrastructure increased that anxiety.

The situation was particularly acute for Sam Ginn and Pacific Telesis, as the demands for capital coincided with the need to pour cash into cellular systems abroad. A smaller, more entrepreneurial company might have cut back or eliminated its dividend. But Pacific Telesis had inherited AT&T's shareholders, the vast majority of whom held their stock because it was a safe, income-producing investment. They expected a dividend every year. Indeed, they expected it to rise. For the first time since divestiture, the company had to increase its debt. Ginn and his CFO John Hulse attempted to cover the capital needs of all the company's businesses, but tensions over the financial strategy steadily mounted. On the international front it became apparent that PacTel might not be able to finance the German and British cellular systems at the same time. At home, a one-time write-off from the failed real-estate subsidiary in 1990 wiped out a substantial portion of the company's earnings. The following year was not much better. Nearing the end of 1991, Pacific Telesis's stock price was down nearly 20 percent, and things didn't look much better for 1992. At one point, a frustrated Hulse told Ginn, "You aren't going to work forever. When people look back and say, 'What was Sam Ginn's legacy?' I'd like for them to have something positive to say other than you had flat earnings for seven or eight years."[38] At the moment Ginn didn't have an answer. When he did, it would stun Hulse and everyone else in the telephone business.

FOURTEEN

The Topaz Solution

In October 1991, when Sam Ginn took the Pacific Telesis Board on a retreat to review the firm's strategy, he was worried. In the short term, his strategy of diversification wasn't producing the kind of results the stockholders and the Board wanted. While the local exchange business – Pacific Bell – anticipated a good year, continuing losses in the PacTel companies would dilute the company's earnings. Because the stock was traded conservatively, the company's share price was likely to go down. This was not good news for any board to hear, especially because Ginn and Sarin had told a similar story a year earlier. The investments Pacific Telesis was making would bring handsome returns, they had explained, but not next quarter. Three more quarters had passed since that meeting. But the forecast hadn't changed.

As he headed for the meeting at Rancho Valencia, along the southern California coast, Ginn was not sure how patient the Board would be. The resort, which was nestled in the dry rolling hills north of San Diego, offered a luxurious California-style site, with clay tile roofs, whitewashed beams, and glazed tile–bordered fireplaces. But the luxury would do little to soften the bad news. The recession in California was deepening, along with the mood of investors, and Sam knew his Board was uneasy.

Ginn and his chief strategist Bob Barada talked to the Board about the opportunities ahead, but they also described the growing divergence between the needs of Pacific Bell and those of the PacTel companies. Lee Cox told the Board that privatization efforts would offer an extraordinary three-year window of opportunity to win cellular licenses in Europe, and when the window closed, it wouldn't open again. Phil Quigley focused on the new regulatory framework and the prospect that someday soon his organization, Pacific Bell, would be able to deliver cable TV and information services over fiber-optic cables to most businesses and homes in California. But to realize this breakthrough – and offer one provider for a host of services – Pacific Bell would essentially have to rewire the state. The Board members listened carefully.

Then it was John Hulse's turn. As chief financial officer, Hulse had agreed to the annual plan reluctantly. Compared to the previous year, it projected a substantial increase in debt and little good news for shareholders, who wanted increased earnings and higher dividends now. To meet its capital needs, Hulse said, Pacific Telesis would have to borrow nearly $1.2 billion and increase its debt ratio to almost 49 percent.[1] Some of the Board members recalled the dark days of the early 1980s, when the company's bond ratings had plunged. They expressed concern and wanted to know how much debt the company could reasonably carry. Ginn worried about their mounting frustration.

During a break, Mary Metz approached him. Metz was the former president of Mills College, a small, respected women's college in Oakland. "This is the same story we heard last year," she said. "More investment opportunities, but also more dilution and flat earnings." Ginn conceded the point, and he couldn't provide any immediate solution. As Metz walked away, Ginn realized that next year he would have to tell this dreary story again. And probably the year after that as well. Any forceful effort to solve the problem, he told the Board in the next session, would have enormous consequences for the company and might even require a radical departure from the strategies of the past. The Board didn't blink. Any time he was ready, they said, he could come back to them with a major proposal to reshape the company.[2]

Ginn didn't have a specific proposal in mind, but for nearly two years now he had been anticipating the need to restructure Pacific Telesis. He and Sarin had discussed the issue before Sarin moved from his position as the top corporate strategist at Pacific Telesis to take a line job at Pacific Bell. Ginn had also raised the issue with Bob Barada, Sarin's successor, but Ginn concluded the time wasn't right for drastic action. "When we do this," he said, "it's going to have an enormous effect on running the business. We're not ready to do it yet."[3]

After the meeting at Rancho Valencia, Ginn knew that the time had come; however, he approached the transformation in the old Bell style, cautiously and quietly. He didn't want to go to a New York investment banker with his dilemma. The news might leak to Wall Street. "If I go to a small firm in the backwaters of the San Francisco financial community," he thought, "I can probably keep this a secret."

Not long after Rancho Valencia, the Harvard Business School honored Don Guinn with its Statesman of the Year Award. Presiding at the dinner was Robert Smelick, the managing partner of Sterling Payot – a small, San Francisco–based investment banking firm. Smelick had done work for Pacific Telesis for a number of years, and he had the confidence of both Guinn and Ginn. After the dinner, Ginn stopped Smelick to say hello and then asked him to set up a meeting at PacTel's office.[4]

When Smelick arrived the following Friday, Ginn explained the situation. The split between the company's two main businesses was growing wider. The company needed to reconcile the conflicting short-term and long-term goals of Pacific Telesis shareholders. It was absolutely necessary to fund Pacific Bell's improvements, but they also needed to generate equity for the wireless ventures and unlock the value of PacTel's cellular investments, which Ginn and his team estimated at $10 billion.[5]

The capital bind at wireless was already a pressing concern. Tight money had forced the company to sell its stake in the PCN network in the United Kingdom because it simply couldn't afford to build both the German and British systems. Ginn knew that the German license had to be retained. In Germany, PacTel and its partners would be competing against only the lumbering Deutsche Telekom. In Britain, PacTel's consortium would be in a market that was already developed, competing against four or five firms. Ginn's Bell System background made him reluctant to enter a market with that kind of competition, especially since his partnership with British Aerospace was problematical and contentious. PacTel favored building out London first and launching service as fast as possible. British Aerospace wanted to build the network for the whole country first. Pressed to make a move he would later regret, Ginn sold PacTel's British stake.

Unhappy to be constrained in that way, Ginn looked for a way to give PacTel more degrees of freedom. After swearing Smelick to secrecy, Ginn concluded: "I want you to think about how we're structured." He was already thinking the unthinkable, a breakup of the two parts of the company. But he wanted Smelick to look at two key issues. Would a split enhance shareholder value? Would an IPO be able to raise the half a billion dollars Ginn thought he would need to fund wireless?[6]

Smelick and his staff studied these problems through the early part of December and came back to Ginn just after Christmas 1991. Their analysis confirmed much of what Ginn already knew. The internal demand for capital would soon force Pacific Telesis to make an ugly choice. If the company continued expending capital ahead of revenue, it would have to take on enormous debt and risk its high credit rating. Even then it was doubtful that Pacific Telesis could borrow enough capital to fund all of its opportunities. Alternatively, it could choose to curtail its investments in either Pacific Bell or PacTel wireless. Cutting investment in Pacific Bell would inevitably spark a fierce reaction from consumer groups and the CPUC, which would insist that the company honor its responsibility to maintain a state-of-the-art local exchange network. That strategy would also put Pacific Bell dangerously behind in the emerging race to deliver broadband services to telephone consumers. On the other hand, choking investment in wireless meant letting go of once-in-a-lifetime opportunities at home and abroad.

"So what's going to happen?" Smelick asked rhetorically. "Well, you're not going to trash your credit. You're *not* going to *not* make the expenditures on the wireline side, because you've got too much infrastructure to protect. And you're not going to dilute your shareholders with massive stock offerings. So you're going to call up the wireless guys and say, 'Don't spend any more money, no matter how good the opportunities are.' "[7] It would be the British experience all over again.

To avoid that dreary choice, Smelick said the firm should consider several options. Telesis could raise capital by selling minority shares in its wireless subsidiary; it could also explore the idea of a tracking stock. Yet these were only stopgap solutions, Smelick said, because Pacific Telesis faced an even more fundamental tension in the long run. As the cost of owning and using a cell phone continued to decline and as companies like PacTel expanded their marketing programs to reach a broader cross-section of the population, a mass market for cellular service would emerge.[8] Down the road, PacTel's wireless ventures would thus challenge Pacific Bell's wireline network in the voice communications market, making the choices about capital investment even more dicey.[9]

As early as 1985, Craig McCaw had predicted this quandary when he complained about PacTel's acquisition of a nonwireline license in the Bay Area. Faced with the prospect that wireless could cannibalize existing wireline traffic, McCaw argued, PacTel and the other Baby Bells would inevitably stifle wireless growth. At the time, PacTel told the courts and the FCC that this would never happen. But now it already *had* happened, and Sam Ginn needed a way out of this bind. Smelick concluded that shareholders shouldn't be caught with one foot in the wireline boat and another in the cellular skiff. They, and PacTel's managers, would do better if the two businesses were divided into separate and independent companies, each free to compete wholeheartedly for the customer's business.

As he stood at this crossroads, Ginn started to muster support for a sharp break with the past by bringing more of his top officers, including his general counsel, into the conversation. His top lawyer pointed out that, if the wireless companies were spun off, then they would probably be free of Judge Greene and the restrictions imposed by the consent decree – restrictions that still governed all the organizations broken out of the Bell System. That news inspired Ginn. Escaping the MFJ hadn't been his intention, but it was a great plus. He was now convinced that it was time to take the question of separation forward and promptly called a meeting of his top officers.

When Lee Cox got the agenda for the meeting, he noted a single item called "organizational issues." Curious about this vague label, he called Barada to find out what this was about. "It's nothing," Barada responded. Cox got a chance to find out what "nothing" meant after Ginn took them for a walk up Sutter Street. "For several weeks," Ginn explained, "Bob Smelick has been looking at our organizational structure, and he has a proposal that we need to take seriously." Cox looked across the table at Phil Quigley as if to ask if he knew anything about this. Quigley registered his own surprise.[10]

When the meeting reconvened in the Sterling Payot offices, Smelick laid out the proposal to split the company. Cox accepted the logic of separation immediately, but as he knew, one big question loomed on the horizon. What would be the reaction of the CPUC? Would they try to stop a split? – probably. But the team was so convinced of the merits of the idea that they had to take it seriously. Ginn gave the go-ahead for a major study of all aspects of the proposal.[11]

The secret study was known as Project Topaz. Over the next four months, Ginn, his colleagues, and Sterling Payot developed a range of scenarios that included a spin of either Pacific Bell or the PacTel diversified companies.[12] Ginn met with Board members in a series of dinners that focused on the issue. "My initial reaction," said one Board member, "is to say we shouldn't do this." Ginn responded by pointing to a growing body of research showing that deconglomeration added value for shareholders, eliminated agency problems, reduced negative synergies, and in most cases resulted in more focused management.[13]

Smelick reiterated these points when the full Board met in April 1992. He also described the diverging strategies and characteristics of the wireless and wireline businesses, highlighting the regulatory and financial burdens imposed on the wireless companies by their connection with the local exchange business. He talked about the differences between the more financially aggressive investors in the wireless sector and the shareholders of the Baby Bells. He also summarized the problem of conglomeration. " 'Pure plays'," he said, "are valued by the market at a premium." The wireless operations and the local exchange company would be worth more apart than they were together.

Following Smelick's presentation, Ginn told the Board that he and his management team believed the separation proposal made sense. The company should commit to a formal study of the concept, and that meant telling shareholders and the world that a spin was a serious possibility. As Ginn reminded the Board, they had no time to waste. The pace of consolidation in the domestic cellular industry was increasing. Wireless opportunities abroad were already on the table. The company had just won a license in Portugal, and it was preparing applications for Japan, Italy, and Greece. The investment window was open for companies willing to be bold and able to raise sufficient capital. If they were going to adopt the Topaz solution, it was time to move – quickly!

Looking to Europe, Ginn could provide an analogy. Racal, the British defense contractor, had launched its wireless subsidiary Vodafone in 1985. Frustrated by the market's unwillingness to give full value to this rapidly growing business and worried about a possible takeover, Racal had floated half of Vodafone's stock in 1990. But while this move gave Racal a temporary boost, it didn't give investors the pure-play opportunity they wanted. The share prices of both Vodafone and Racal began to drop again. Racal's management finally concluded that only by spinning off Vodafone could they realize the value of the wireless assets. They were right. When the company was divested in September 1991, Vodafone's shares soared.[14] Ginn thought that Pacific Telesis would enjoy a similar experience. Convinced by management's arguments, the Board agreed to a formal exploration of the separation proposal.

Initially, Ginn hoped to complete the study and have a decision by late June. But as the enormity of the task became clear and the issues increasingly tangled, the deadline slipped from June to August. Then some new information made the Board uneasy. In order to provide added due diligence and objective counsel, the Board had hired the firm of J. P. Morgan to offer a second opinion. Morgan came back in July expressing concerns. Sprint's lowball offer of $3 billion in stock for the cellular company Centel (in May) had sent cellular share prices into a tailspin.[15] Adding to the uncertainty, a federal government report that month suggested that cellular companies were charging too much because of a lack of competition.[16] The CPUC was working on a study that was likely to echo that critique, and Morgan concluded that the political and economic climates were likely to keep cellular prices depressed for some time. Thus, even a separate PacTel wireless company might not be able to raise enough equity to meet its capital needs. Down the road, Pacific Bell might also face similar problems. Morgan's report unnerved the Board, leaving the Topaz strategy in doubt.[17]

The Board was also concerned about the tradeoff between lost opportunities and potential synergies. The announcement of the spin study had sparked criticism from a number of industry analysts, who argued that the future of telecommunications was in convergence. According to this view, the markets for telecommunications, newspapers, and cable TV were coming together, offering economies of scope and scale to service providers. In the near future, customers would embrace the first carrier that offered a single package of wireline telephone services, cable television, wireless telephony, and – in a pre–World Wide Web world – electronic information services.

Ginn understood this argument. To hedge his bets, he had already guided Pacific Telesis into the cable television business on a small scale. But a variety of consultants and speakers who came to talk to the Board also made it clear that regulatory impediments to synergies, at both the state and federal levels, would not soon evaporate.[18] Professors Robert Harris and Anthony Oettinger from U.C. Berkeley and Harvard made it clear that the Board should not base its decision on the assumption that there would be convergence. Telecommunications guru Peter Huber confirmed this view. Although the MFJ restrictions were destined to be removed, Huber said, he anticipated that it would not happen before the end of the decade.[19]

Of all the groups that studied the proposed spin, Ginn's own management team supplied the most conservative perspective. They predicted that the net result of the spin would be positive for shareholders, but they cautioned that separation was not likely to create the bump in market value that many other companies had enjoyed. The spin would not "disadvantage" either company, the report concluded, and it might make it easier for the two companies to work together as joint venture partners in some markets. But the management team promised no rosy returns.[20]

Ginn's thoughts about the costs and benefits of a spin were influenced by the prospect of a major technological shift in cellular service. To promote competition, the FCC was considering allocating spectrum for a new cellular service called personal communications services (PCS). Like PCN in the United Kingdom, PCS offered the possibility of a low-cost system of lightweight portable phones linked by low-powered microcell transmitters. Because of the lower cost structure and the lighter handsets, analysts predicted that PCS would attract 50–60 million users by the year 2002 and would be a $660-billion business by 2010.[21] Some suggested that, in addition to being competitive with existing cellular systems, PCS would encourage many callers to bypass the local wireline loop altogether.[22] Thus this new technology loomed as a double-barreled shotgun aimed at the Baby Bells' largest revenue source – the local exchange – and their fastest growing new business – cellular.

The FCC in 1992 was wrestling with the question of how to license PCS. A number of industry groups, including cable TV providers, were pushing the government to grant them licenses and keep existing cellular firms out of this new round of spectrum allocations. Ginn, like other cellular leaders, was eager to expand his company's geographic footprint by capturing some of the PCS business. At the end of 1992, PacTel Cellular was the fifth largest firm in the United States in terms of cellular subscribers and the sixth largest in terms of total population coverage. It trailed only Bell Atlantic, Southwestern Bell, BellSouth, GTE Mobilnet, and McCaw Cellular, which led the pack.[23] PacTel's management thought that the government was not likely to allow companies to apply for PCS licenses in markets where they already provided cellular service. Thus, the best licenses in California would go to others. If the PacTel Companies were spun off, however, Pacific Bell would be able to compete for these licenses and shareholders would benefit.[24]

While the Board considered the options, Ginn scoured the globe in search of a strategic investor who could provide $100 million in equity and so minimize the uncertainties of an initial public offering. He met Werner Dieter in London and asked if Mannesmann would be interested in owning 10–15 percent of PacTel wireless. But Dieter had capital concerns of his own. After pouring more than a billion dollars into the construction of the D2 system, Mannesmann was only just beginning to earn some revenue, and the build-out of the system was still a long way from complete. Cell sites were being added. This capital drain and the Germans' uncertainty about the future of cellular were creating problems for Dieter. On the question of investing in PacTel, Dieter told Ginn, "I don't think I could get my Board to support it."[25]

With Mannesmann out of the picture, Ginn looked to Asia. Singapore had a huge pension fund whose managers were interested in making some strategic investments. Two or three times Ginn flew to this southeast Asian city-state trying to make a deal. Many of the top-level pension managers he met had been educated in the United States, and they understood the potential of the American market. The managers and Sam almost completed an agreement, but then the pension team insisted on a substantial discount on the IPO shares. Ginn was unwilling to concede this much and the deal fell apart. Disappointed, he returned to the pencil pushers in the investment banking community and asked them to reevaluate PacTel's potential in the equity markets. The Pacific Telesis Board was using Shearson Lehman and Salomon Brothers to develop a similar calculation of the company's potential. In contrast to J. P. Morgan, they predicted that a PacTel IPO, despite the depressed state of cellular shares, would be "an event offering" that would attract enormous interest among investors.[26]

To the surprise of J. P. Morgan and Sam Ginn, however, the cellular "event" that followed did not involve Pacific Telesis. Burdened by enormous debt, the drop in cellular stock values, and the fact that his company had still not produced an operating profit, Craig McCaw had begun to look for a back door. For years, he had contemplated an alliance with AT&T that would give him access to Ma Bell's brand name and capital. What he had to offer, he thought, was a way for AT&T to bypass the Baby Bells in the local exchanges. Cautiously, McCaw had approached AT&T in the summer of 1990. The timing was right. With cellular industry revenues increasing by 40 percent a year and the total number of subscribers in the United States approaching the 10-million mark, AT&T could no longer afford to ignore this competition. But true to its heritage, AT&T moved slowly. Talks dragged on for two years, and McCaw began to consider other options, including a deal with BellSouth. Unable to obtain desirable terms there, he looked again to AT&T. He now realized that AT&T would never make a major investment without an opportunity to take control. To realize the potential of his business, he reconciled himself to the fact that he would have to let it go. He and AT&T's CEO Robert Allen finally struck a deal in August 1992: AT&T would buy one third of McCaw Cellular for $3.8 billion, along with an option to buy the company outright.[27]

This sent a shock wave through the industry in November 1992, when the acquisition was announced. Ginn saw it as a mixed blessing. The bad news stemmed from the threat that AT&T would use McCaw's system to bypass the local exchange, a threat that sent Baby Bell stocks falling sharply. The good news came from the market for cellular shares. The deal sparked a surge in cellular share prices and removed a great deal of the uncertainty surrounding a PacTel IPO.[28] Analysts now told Ginn that he would have no trouble raising $600 million or more. This was what Ginn needed to fund European expansion.

Ginn was now absolutely certain that a spin was the right course of action. The decision had everything to do with the explosive growth of the wireless world. "Our industry is on the threshold of a revolution," he wrote the Board in October. For the local exchange companies, the revolution would be delayed by the painful process of regulatory change, but a technological transformation was inevitable. In wireless, "the revolution has already started," he said. "Margaret Thatcher started it all with the privatization of British Telecom and the creation of competitive alternatives. The move away from government ownership is spreading globally." Freeing the wireless company from Pacific's current regulatory and financial situations "would immediately release a plethora of creativity among planners ... and the value will be enormous." In closing, Ginn urged the Board to make a decisive move forward. "We are not facing a decision between disaster and separation," he wrote. "The management team and I will forge ahead with whichever decision you make." Nevertheless, Ginn made it clear that he would vote for a spin.[29]

Confident in the wake of the McCaw–AT&T deal that the IPO would be successful, the Board gave its unanimous approval to the plan. The decision captured the headlines of the business press in San Francisco and on Wall Street the next morning. As Ginn told reporters, the split "produces two strong companies, each with an exciting future as we approach the 21st century."[30] With Pacific and Nevada Bell generating the bulk of its revenues, Pacific Telesis Group would remain one of the nation's largest local exchange carriers. The spin-off company, valued by some analysts at $4–$5 billion, would emerge as one of the world's largest wireless providers.

In discussing the spin, everyone speculated about Ginn's personal future. Craig McCaw's long-time right-hand man, Wayne Perry, was convinced that he knew a Bell-head when he saw one. He predicted that Ginn would keep his comfortable job with Pacific Telesis, and he bet billionaire Craig McCaw five dollars that Ginn would stay. McCaw, who had seen the entrepreneurial light in Ginn's eyes, took the wager.[31]

Ginn already knew what he wanted to do. After a 32-year career in the Bell System, he was ready for a change. He had grown tired of the claustrophobic world of regulators, labor unions, and the worst aspects of the Bell bureaucracy. He was eager for an opportunity to build an organization around values rather than rules. And he was also not blind to the enormous personal fortunes being made by Craig McCaw and others in the cellular industry. He was, however, loath to make his desires known until he was confident that Pacific Telesis had the resources and a management team in place that could carry the company through the competitive era ahead. The memory of Charlie Brown's leadership during the breakup of the Bell System was still fresh in his mind. Fortunately, the Bell System had conditioned him for years to focus on grooming his own successor, and he had not abandoned that part of his cultural heritage.

Having led Pacific Bell for five years, Phil Quigley was ready to take Ginn's job. He fully understood the mix of regulatory politics and competitive forces that would shape Pacific's future. Given his experience in launching PacTel Cellular, he was also uniquely qualified to prepare the firm for entry into the PCS land rush. As Ginn discussed the issue of leadership with the Board, it became apparent that they agreed. It was also obvious that Ginn's presence at the head of the spin-off would reassure Wall Street. Hence, with their decision to approve the spin, the Board also approved Ginn's appointment as CEO of the new wireless company and Quigley's promotion to CEO of Pacific Telesis.

Although Ginn and the Board agreed that the Topaz strategy was reasonable and prudent, there were others who saw things differently. Over the next twelve months, some of Sam's critics accused him of being a hardball, self-dealing corporate chieftain intent on swindling California's ratepayers. Others, the market analysts, questioned the spin. Back in Anniston, Alabama, Myra – Ginn's aging but intrepid mother – asked her son if he knew what he was doing. She was concerned about how she could explain to her bridge partners that her son was no longer the CEO of a Fortune 50 company. Of more concern were the critics headquartered a mile or so away in the offices of the California Public Utilities Commission on Van Ness Avenue. There, a battle continued between the champions of deregulation and those members of the staff who believed it was their duty to protect California's historic role as the nation's leading advocate of consumer protection and aggressive utility regulation. In their view, the Commission had not been created to foster innovation or to approve of restructuring that moved assets out of their domain. Pacific Telesis had to deal with both sides in this internal conflict, but Ginn firmly believed it was time to draw a line in the sand and challenge the Commission to embrace the Topaz strategy and thus an entrepreneurial vision of the future.

FIFTEEN

Riptides of Reform

Within hours of the Board's decision to consider the spin in April 1992, Sam Ginn rode up Market Street to talk with Dan Fessler, president of the CPUC. Ginn set up his pitch to Fessler by briefly reflecting on the past.

"You know, Dan," he started, "I know you have been terribly concerned about this whole issue of cross-subsidy. And you have put in all these rules so that I can't transfer people from Bell to the PacTel companies without paying this big fee, but if I send one from the diversified companies back to Bell there's no charge."

"Yes, that's right," Fessler responded, his demeanor changing slightly as he prepared for the argument.

"Well, I've decided to solve that problem," Ginn said with a poker face. "I've decided to look at spinning off a separate company, so you won't have to worry about cross-subsidy ever again."

Ginn remembers that Fessler's face froze. "You mean a completely separate company?" he asked.

"I mean a completely separate company," Ginn continued, "with a different board of directors and different shares."

Fessler saw the implication of a spin. PacTel's profits would not be available to subsidize basic phone service if Ginn made good on this plan. To fight him, Fessler and his staff would have to contradict the basic arguments they had made about cross-subsidies and transfer pricing. As Ginn well knew, Fessler was not happy when he left.[1]

The mood at the Commission became even darker when Pacific Telesis filed its official notification of the spin in January 1993. Testing the limits of regulatory authority while trying to avoid direct confrontation, Pacific Telesis maintained that neither the FCC nor the CPUC had the authority

CPUC headquarters, Sacramento – The imposing façade of the new headquarters of the California Public Utilities Commission on Van Ness Avenue in San Francisco evokes the regulators' long and powerful rule over the state's utility companies.

to block the separation because the spin would not change the regulatory relationships between the state and the two separated companies. Both would still be subject to regulation. There was no transfer of ownership, either, because current shareholders of Pacific Telesis would end up as the owners of both companies after the spin. The company claimed that it did not need the regulators' permission.

The Commission's consumer advocates vigorously disagreed. In a brief filed with the Commission, the DRA (Division of Ratepayer Advocates) argued that the spin constituted a transfer of ownership and was therefore subject to CPUC approval. If the CPUC determined that this change was not in the public interest, they asserted, then regulators could block the deal. Several labor and community groups joined the DRA in claiming that ratepayers should be compensated if the wireless business was spun off as a separate company.[2]

The Commission stepped into the fracas by opening an investigation in February. After preliminary hearings, Pacific Telesis asked for an expedited ruling. But the Commission moved slowly. At formal hearings before an administrative law judge in July, the battle lines were drawn. According to the DRA, Pacific Telesis had built up the wireless company by putting ratepayer assets at risk. Ratepayers should be compensated at market prices for the loss of those assets. Consumer groups supported the DRA, asserting that Pacific Telesis should pay as much $2 billion to compensate ratepayers for the money invested in cellular. The DRA also charged that Pacific Telesis had starved Pacific Bell for capital, squeezing $1.17 billion out of the company in depreciation charges over and above its new capital investments in the late 1980s in order to fund the expansion of wireless. This "disinvestment" would come back to haunt Pacific Bell's ratepayers after the spin, the DRA said. Without this capital, Pacific Bell would be forced to ask for a rate hike to finance the rewiring of California, and ratepayers would have to pick up the tab.[3]

Pacific Telesis disputed every aspect of this analysis. The DRA, the company said, was comparing apples to oranges. Given the massive investments the company had made in new switching systems and infrastructure in the 1980s and given the declining cost of telephone equipment, it could only be expected that reinvestment would not match depreciation. The company pointed out that the DRA and others had recently blasted the company for spending too much on the modernization of its network, accusing the firm of needlessly increasing the size of the rate base.[4] The CPUC had accepted that argument and penalized Pacific Bell. The regulators, Pacific Telesis said, couldn't have it both ways.[5]

The DRA also claimed that the PacTel name was a ratepayer asset. If the new wireless company intended to use it, then ratepayers should be compensated with a licensing fee equal to 3 percent of the firm's gross revenues. With gross revenues of approximately $900 million in 1992, this licensing fee would have been worth $27 million a year to ratepayers.[6] Because income was expected to rise dramatically in the near term as revenues from Germany and other new systems came in, Pacific Telesis had a powerful incentive to oppose this fee. As Pacific Telesis pointed out, its name belonged to the holding company and not to Pacific Bell or its ratepayers. Since divestiture, the holding company, not Pacific Bell, had spent nearly $90 million promoting and protecting the Pacific Telesis or PacTel name. It already had the right to license the use of the name as it saw fit.

Among all of the issues debated before the Commission, the most contentious matter involved the early development of cellular technology and the FCC's initial award of licenses to AT&T's local exchange companies – an issue that still aggravated Craig McCaw. The DRA argued that ratepayers had funded AT&T's research. Since ratepayers were put at risk for this money, they should be entitled to a "significant portion of the anticipated gain" in the value of the cellular licenses in California. According to the DRA, the FCC had awarded these licenses to Pacific Telesis without charge because it was a local exchange company. The licenses were, in this view, ratepayer assets – and they were enormously valuable. Indeed, from the DRA's point of view, much of the value inherent in PacTel's wireless operations was not in the towers and switching center the company had installed but rather in the spectrum rights it owned. That spectrum, the DRA said, had been granted to the local exchange company.[7]

Telesis countered that ratepayers had never been "at risk" for funds channeled to AT&T for research and development in telecommunications, nor should they be rewarded for Bell Labs' innovations. Under the terms of a 1956 consent decree, the government had compelled AT&T to license many of its patents to anyone who wanted to make use of them. If the DRA's argument were accepted then ratepayers should have received a return on all Bell Labs inventions, including the transistor, but neither the CPUC nor any other state agency had asked to share that revenue. As Sam Ginn had pointed out to Fessler, the CPUC and the FCC had mandated that cellular companies operate as fully separated subsidiaries, and the CPUC had imposed stringent rules on affiliate transactions to ensure that the cellular company didn't leverage Pacific Bell assets. Since the assets of the California cellular properties would remain in public service and be subject to CPUC regulation after the spin, Pacific Telesis said, the ratepayers would not be deprived of any value transferred by the FCC in its original license allocation.[8]

After the hearings ended in July, the wrangling continued in the press. Ginn now began to worry that he might never be able to complete the separation. Remembering the game of chicken the CPUC had played with AT&T before divestiture, he had every reason to worry. When AT&T decided to acquire McCaw outright in August, the pressure to move quickly increased. Now there was a great deal more to worry about. With its deep pockets and brand name – and with McCaw's cellular experience and expertise – AT&T suddenly loomed as a major rival in both the domestic and international markets. The McCaw deal also boosted cellular share prices again, making it even more important to get the wireless IPO to Wall Street while the market was hot. Ginn's team now calculated that a 10-percent offering of the wireless shares would net $1.2 billion in cash, an impressive war chest for the international campaign ahead.

Then, just as Ginn and his team were getting ready to test the market, the CPUC's administrative law judge threatened to sink Ginn's plan. On the issue of jurisdiction, the judge was unequivocal. The CPUC had authority to approve or block the transaction. Therefore, the judge recommended, the CPUC should order Pacific Telesis to submit a formal application, and the regulators should begin a full-blown investigation into a variety of issues relating to the effect the transaction would have on Pacific Bell and California ratepayers. On the issue of Pacific Bell, the judge suggested that future wireless earnings should provide a source of equity for the regulated firm. In regard to PacTel's name, the judge said that ratepayers were entitled to compensation.[9]

The judge's recommendations confirmed all of Sam Ginn's deepest fears. Repeating the pattern of the AT&T years, the CPUC intended to squeeze out of the holding company the capital needed to modernize Pacific Bell's network. Only this time, instead of leveraging the profits from Illinois or New York telephone customers, California regulators would tap the income from cellular operations around the globe. While other cellular entrepreneurs raced to win licenses in other countries, PacTel would be bogged down in CPUC proceedings for at least another year. If the judge's conclusions held up, Pacific Telesis might have to pay $1 billion or more to complete the separation.[10]

Facing delay and what Pacific Telesis termed a "ransom," Ginn blew up. He sent a voice-mail to all of the company's employees castigating the decision. "The Staff Proposal is so far off the mark that you have to wonder how such a decision could come out of the CPUC process .... This opinion is one more example of how regulation is simply not keeping pace with the rapid changes in this industry – particularly in California."[11] Ginn carried his complaints to the public. In a letter to the editor of the *San Francisco Examiner,* he said that Pacific Bell customers did not have an ownership interest in Pacific Telesis, "any more than purchasing a Buick gives a buyer an interest in General Motors."[12]

The company itself launched an aggressive counterattack. A week after the CPUC issued the proposed decision, Pacific Telesis hand-delivered a response directly to the commissioners. The document characterized the proposed decision as "fatally flawed," "unbalanced and extreme." According to the company, the judge had simply redrafted the DRA's briefs and the contentions of other opponents of the spin. The judge had not relied on evidence in the record. Moreover, he had violated agreements regarding confidential information. If the Commission ratified the proposed decision, Ginn's lawyers asserted, it would derail the spin.[13]

Pacific Telesis's critics cried foul when they learned about this direct response to the commissioners. It violated California's sunshine rules regarding *ex parte* communications, they argued, and was consistent with a pattern of initiatives by Pacific Telesis designed to short-circuit the regulatory rules.[14] They asked the Commission to impose sanctions on the company for breaking the rules.

Beset by a hornet's nest of regulators, Ginn suddenly found himself in trouble with the investment community as well. Things started to warm up when Bell Atlantic stunned Wall Street with the announcement that it would buy Tele-Communications Inc. (TCI), a cable company, for nearly $33 billion. This put a spotlight on the issue of convergence. Across the country, the other Baby Bells were talking about their interests in cable and reassuring their shareholders that they, too, were preparing for the day when local exchange companies would deliver telecommunications and television to customers' homes. They predicted that convergence would generate enormous economies of scope and additional profits for shareholders.[15]

Critics berated Ginn for continuing to pursue the spin. "Some analysts argue that PacTel's singular focus on the spin-off over the last two years," wrote the *Los Angeles Times,* "has allowed the company to slip behind in the race to bring multimedia services to the doorsteps of Californians."[16] Like a storm moving across open water, opposition to the spin-off was gaining force. Ginn knew he had to have a decision soon, or else two years of effort would be wasted and the Alexander Graham Bell opportunity would remain just beyond his grasp. For Ginn, the last hope lay with the commissioners themselves.

Commissioner Norman Shumway had anticipated staff resistance to the spin. A lawyer and a conservative former Congressman from Stockton, Shumway had been appointed to the CPUC by Republican Governor Pete Wilson in February 1991. He was a fierce opponent of big government. Seeking to take control of the spin issue, he told the press that "a delayed decision in this case may well be tantamount to a denial of the proposed spin-off." The CPUC, he said, did not need more proceedings. It had enough information to make a decision. "The citizens of California would surely be victimized," he continued, "if the matter were so prolonged that the opportunity it presents for economic growth and development to our state would be lost."[17] Shumway announced that he would submit his own proposed decision to the Commission as an alternative to the judge's draft.

Shumway had brought up a sensitive issue. Economic development was PacTel's ace in the hole. With California still suffering from a prolonged recession, state officials were sensitive to the accusation that the Golden State was not a friend to commerce. Ginn told reporters, "This is not the time to tell the world that regulation makes it too tough to do business in California." To support his case, Ginn lined up top officials in the Wilson administration. He also met with the state's leading Democrat and power-broker, Assembly Speaker Willie Brown, to argue that his plan would help California's economy. Although he knew Ginn was a Republican, Brown accepted his logic and agreed to author an opinion page column in the *Los Angeles Times* supporting the separation.[18] Brown's public support was critical. It neutralized Democratic opposition and prevented the fight from becoming a partisan issue. Long experience with the regulatory environment had made Ginn and PacTel adept at politics in California and in Washington, and in the late summer of 1993 they and the cellular industry won a victory on Capitol Hill that increased pressure on the CPUC.

Congress transformed the regulatory environment for wireless in the Budget Act of 1993. It took away the ability of state regulators to control entry into the wireless market and removed their authority to regulate rates. The federal government dramatically reduced the ability of state commissions to sustain a command-and-control regime and gave competitors less opportunity to control entry by working the political system to their advantage. Cooperating with the cellular trade association, Ginn had lobbied for these changes while the CPUC lobbied against them. The CPUC quickly sought and was promptly denied FCC permission to be exempt from the new legislation.[19]

With their wireless regulatory authority under pressure, at least two of the CPUC's commissioners sought a compromise with Pacific Telesis. The administrative law judge's conclusions seemed to all but foreclose the possibility of any deal that would make sense for the company. But at the urging of two of the commissioners, Ginn's staff sat down with their opponents to try to negotiate a settlement. Those talks collapsed in the last week of October over the price that the DRA and others wanted for compensation.

Meanwhile, the commissioners themselves repeatedly postponed a vote on the issue. Transcending conventional political ideologies, the five members of the CPUC – all appointed by Republican governors – were split into two factions. Commissioners Daniel Fessler and Gregory Conlon sided with the consumer advocates, who argued that PacTel should compensate ratepayers for cellular spectrum. Commissioners Shumway and Patricia Eckert thought that the spectrum belonged to shareholders and that the company should be allowed to proceed with the spin. Divided down the middle, the commissioners postponed scheduled votes on the issue three times in six weeks.

The fate of the spin came down to one vote cast by the Commission's newest member, Jessie J. Knight, Jr. A 42-year-old African-American marketing executive from the San Francisco Newspaper Agency, Knight had stepped into this battle in August. Having spent most of his career on the corporate fast track at Dole Foods and then in the newspaper business, he knew little about utility regulation and had to hustle to come up to speed on the PacTel issue. In November he reached a decision.

Many people anticipated that on November 2 the Commission would once again postpone a vote. As Patricia Eckert and Jessie Knight rode the elevator down to the hearing room, they discussed the issue. A lawyer and a champion of free markets, Eckert was the senior member of the Commission and the last remaining appointee of Republican Governor George Deukmejian. In the elevator, she reiterated her key points: that the Commission was trying to move away from command-and-control regulation, that competition was the way of the future, and that the wireless business belonged to Pacific Telesis shareholders and not to ratepayers. She stressed the opportunities for California to take a lead in the global wireless industry. While expressing reservations about how the assets of Pacific Telesis would be divided, Knight conceded that the Commission had to show that it could act decisively. The political and economic pressures were too great, he said, to allow further delay. In that moment, California moved one step further along the path blazed by Margaret Thatcher and other advocates of deregulation.[20]

Knight's vote enabled the Commission to make a ruling remarkably favorable to PacTel. The Commission approved Pacific Telesis's plan without requiring the enormous payment advocated by the DRA. Ratepayers, the Commission said, did not deserve royalties on the PacTel name. The Commission also agreed that all Pacific Telesis investments in the wireless business had been properly made with shareholder profits to which ratepayers had no claim, and the Commission concluded that ratepayers should not share in the appreciation of the cellular business. Only with regard to the issue of cellular research and development did the Commission side with the DRA. The agency ruled that on this matter, Pacific Telesis should reimburse ratepayers $7.9 million (with interest, the figure came to $48.7 million).[21]

After the decision was announced, Eckert explained to the press that California regulators, like businesses, needed to keep up with the fast pace of change in telecommunications. Knight hoped the vote would send a signal "that California is a good place to come and do business and to grow." The two dissenting commissioners said they did not oppose the spin per se, but they still believed that PacTel owed compensation to ratepayers. They would be watching, CPUC President Dan Fessler said, to see if Sam Ginn's "rosy predictions" for the California economy came true.[22]

At Pacific Telesis Group headquarters at 130 Kearny Street, Ginn took a moment to enjoy his victory before he started to worry about PacTel's next move. "We don't have the order yet," he told reporters, "but we're satisfied to be at a point where we can choose our destiny." Once again he reiterated PacTel's mission: "We're creating an international wireless telecommunications company with specialized expertise based right here in California." The wireless company, he said, would "help keep California at the forefront of this rapidly developing industry."[23] Though he had convinced the commissioners on Van Ness Avenue to open a path to the future, Ginn knew that he and his wireless colleagues still had much to do if they were going to win hearts and open wallets on Wall Street.

SIXTEEN

AirTouch

Throughout the regulatory battles, Pacific Telesis had been struggling to convince the investment community that the spin-off was a good idea. One analyst for Goldman, Sachs told the *Wall Street Journal,* "This leaves Pacific Bell very vulnerable." Another analyst for Paine Webber thought the wireless business would regret losing its connection to the Pacific Bell cash cow. "This makes no sense," he said. "No Bell company or GTE is going to follow Pacific Telesis's lead."[1]

While he couldn't ignore these criticisms, Ginn was convinced that he could change Wall Street's point of view. Time was short. Pacific Telesis had hoped to have an initial public offering of wireless shares within six months of the Board's decision, while cellular share prices were still high. They nervously watched the international arena, hoping they could get enough cash into the bank before the day arrived when they would have to turn down or sell a license simply because they couldn't afford to build another system. But as they waited on the CPUC in 1993, they were forced to delay the stock offering, and then to delay it again.

News of the AT&T–McCaw deal had, however, jarred Ginn and his co-workers into action. Late in August, before they had even received the administrative law judge's proposed ruling, the company filed its registration statement with the Securities and Exchange Commission for an IPO. Hoping the CPUC would reach a decision soon, they laid the groundwork for an IPO of 10 percent of the wireless stock.[2] When the Commission gave them the green light in November, Pacific Telesis quickly scheduled the sale for December 3. At a proposed price of $21 a share, the offering would bankroll Ginn with $1.2 billion.

The tight deadline gave Ginn and his group only a month to tour the world's financial centers and make their case. Because they intended to focus on growth and not immediate profits, their strategy was to concentrate on investors and institutions that were likely to buy and hold the stock. To cover the cities and markets they needed to hit before the third of December, Ginn's senior staff divided into three teams.[3]

They all emphasized that the new company was a global enterprise. Beginning and ending their road show in New York, the teams traveled to 27 cities in the United States, plus Montreal and Toronto, 13 cities in the United Kingdom and Europe, as well as Hong Kong and Tokyo – all in three weeks. At one point, Ginn and the company's treasurer, Mohan Gyani, made their pitch to nine different audiences in one day. Jan Neels and CFO Chris Christensen had similar experiences. One day, they started with a breakfast presentation in Rotterdam, ate lunch in Paris, gave an afternoon talk in Milan, and spent the night in Frankfurt. Rushing to stay on schedule, they were nearly arrested when their corporate jet pulled into a parking slot between planes belonging to the Queen of Sweden and the President of Germany. By the end of the three weeks, Arun Sarin and Lee Cox had repeated their sales talk sixty times.

At all of these meetings, the team hammered away at the same message. With more than 1.1 million cellular and 1.2 million paging customers, their new company was already among the top five wireless companies in the United States. Los Angeles and Southern California accounted for nearly 40 percent of the cellular business, but the company was diversifying geographically. Investments in licenses in Georgia, Michigan, Ohio, Missouri, and Kansas accounted for another 40 percent of the company's cellular subscribers.[4] They were poised for substantial growth overseas, with systems in Germany, Portugal, Spain, Sweden, Japan, South Korea, and Thailand either up and running or about to come on line. Already the company had more than 250,000 cellular and paging customers abroad. All this business generated plenty of revenue – nearly a billion dollars' worth in 1993 – but not much profit. As Ginn told investors, the profits would come in a couple of years.[5]

The road show was exhausting but successful. On December 3, 1993, Ginn made the first trade in the wireless company's stock on the floor of the New York Stock Exchange. He bought one share each for his three children at the opening price of $25.50 a share.[6] By the end of the day, more than 25 million shares had changed hands. Within a week, the underwriters had exercised an option to purchase an additional 8.5 million shares, raising the total value of the IPO to $1.57 billion. This made it the third largest in the nation's history. The new company was now valued at $12.4 billion – more than Sprint or McCaw, and only $900 million less than the nation's second-largest long-distance carrier, MCI.[7]

For Ginn, the road show had been exciting but also humbling. As CEO of a Fortune 50 company that had inherited the mantle of the Bell System, he was used to being treated with respect. During the road show, however, he was just one more guy looking for money. Analysts showed up late. While he was talking, some of them read the *Wall Street Journal* or ate their breakfast. "You're in an old conference room, and there's no air in there, and they tell you 'some of the guys are going to be a little late. Go ahead and get started.' And there are two people sitting there."[8] The experience was sometimes demoralizing, sometimes invigorating. For Ginn it was a return to his early days in sales, and he was a good salesman. But now the numbers had nine zeros, not just two.

The road show had a humanizing impact on all of the "salesmen." Accustomed to limousines and luxury hotels, Sam and the others sometimes spent the night in a Super 8 or other budget motel.[9] On one occasion, after a full morning of presentations in London, Ginn finally refused to move unless someone found him some food. Always religious about physical exercise, Sam tried to squeeze in runs whenever he could. Early one morning in Zurich he realized that he had lost track of the way back to his hotel. He couldn't even remember the name of the place. In his running shorts and a sweaty T-shirt, he stopped several people to ask for directions, but of course they didn't speak English. Back in the hotel, Ann was sure he had suffered a heart attack. Finally, Ginn found a woman who could speak English. He described the hotel, the trolley tracks in front, and the jewelry store he had noticed when he arrived late the night before. From his description, she identified the hotel and gave him directions. He arrived just in time to take a shower and make it to his next presentation.[10]

Experiences like these catalyzed the new wireless leadership team at the end of 1993. Armed with a fresh supply of cash, they needed only one more thing: a company name. The CPUC process had made it clear that keeping the PacTel name presented problems that would bedevil the new company for some time after the spin. They needed a clean break. In August 1993, Arun Sarin had launched an effort to develop a new name that would accurately convey to customers and partners the company's mission. He solicited names from employees.[11] A team of consultants collected thousands of possibilities that they narrowed down to a couple of dozen to present to Ginn and the other executives. They rejected those that sounded too high-tech. Ginn insisted on having a name that emphasized cellular's connection between people, not the technology.

He liked "AirTouch," which he thought resonated with human contact. It was oriented toward the consumer and was distinctive.[12] As he told the press on the day the name was unveiled, " 'Air' exemplifies freedom, spontaneity, ubiquity ... and 'Touch' is highly personal, suggesting staying in touch. But perhaps most important," he said, "AirTouch is the only name among our major competitors that most clearly suggests a benefit to the customer."[13]

Critics made fun of the name. Some insisted that it sounded more like an air freshener than a global telecommunications enterprise.[14] New Yorkers took the opportunity to dig California: "From the land of the air kiss comes the air touch."[15] This same critic poked fun at the logo – an arc of two elements, one blue and the other yellow – saying it "looks more like the Gateway Arch in St. Louis after Godzilla stepped on it." But Ginn was not dismayed. Over time, he said, "the quality of the service is what makes the brand."

In order to establish the AirTouch brand, Ginn had to figure out how to leverage the best part of the Bell traditions, take advantage of economies of scale, and still remain more flexible and innovative than most of the large bureaucracies in telecommunications. One of the first things he had to change was his own leadership style. For years his mornings had begun like a scene from a Frank Capra film. After an early-morning workout in the gym, he would stride into the Pacific Telesis building and enter the executives' elevator. Dressed in a suit and tie – with all the spit and polish of his Alabama Sunday mornings, his years in ROTC, and the quarter century he had spent in the Bell System – he would walk past the stations of the executive secretaries. As he passed, each one looked up and said, "Good morning, Mr. Ginn," in a subdued, respectful voice. In 1993, when the company was locked in its struggle with the CPUC, he barely said a word. Some of the employees referred to him as "the iceman."

But that had to change. At the wireless company, there was much that could not be done by an iceman. He and his managers had to shape and reshape the organization's operations and culture. Since he had launched the enterprise in 1983, it had already begun to acquire a distinct set of values that blended the engineering discipline of the Bell System with the entrepreneurial spirit of Silicon Valley. But the culture was still young and developing as the company hired new employees and struggled to keep pace with surging demand and new licenses. Most of the workers were members of Generation X, rather than aging Baby Boomers like the employees of Pacific Bell. Many at the wireless firm were not much older than Ginn's own children. He would have to become a teacher as well as a boss.

For months before the spin, the executive team had talked about the kind of culture they hoped to build and sustain. Ginn believed that focus was critical to organizational success. He wanted to keep the company's goals clear and simple – and its performance measurable. Cox was the motivator. He understood how to rally the troops and ignite their passions. Sarin was the strategist. Chris Christensen, the CFO, focused on what the company had to do in order to please Wall Street. As they outlined the future of their new company, the team articulated three basic targets: to be rated number-one by customers in every market; to achieve a world-class employee satisfaction rating; and to double the share price within four years. They knew they could achieve their goals only if the organization had a gut commitment from its employees.

In an effort to win their support, AirTouch gave substantial freedom to employees to shape their own careers. The Bell career paths that Ginn, Cox, and Christensen had followed had been highly structured. In the new company, however, the team eschewed structure, embracing the open offices that characterized Silicon Valley, a more casual dress code, and a walk-around style of management. To keep spirits high, they gave employees a stake in the company's success. In place of the rigid Bell compensation systems, he and his team created a program they called "two by four." On January 1, 1994, they took stock equal to 10 percent of everyone's salary – from top executives down to the mailroom clerk – and put it into a restricted account. If the company's share price actually doubled in four years, then employees would receive that stock at face value. To keep employees focused on the firm's goals, they posted the stock's daily price in work areas. Unions had not acquired a foothold at the wireless operation. Ginn and Lee Cox tried to create an environment in which employees felt they owned the business and didn't need a union to protect them. Both men fully understood that, by not having a union involved, they could keep changing their organization and operations to suit a marketplace that appeared to be unbelievably dynamic.[16] Their Bell experiences suggested that, as long as they could remain nonunion, they would be able to remain extremely flexible all of the way through the organization.

They set out to focus employees on a core set of values that would enable them to respond appropriately when dealing with customers, suppliers, and the public without having to go up the chain of command or read an administrative manual. In a regulated environment, the Bell approach to order and hierarchy had worked very well for many decades.

Ginn appreciated that record of accomplishment. But the wireless company was in a fluid economic setting that called for quick responses and sustained innovation. Elaborate rules and prescriptions – that is, the normal aspects of business bureaucracy – would slow the pace of change. The central managerial problem was to select and preserve the best of the Bell traditions while blending them with the kind of entrepreneurship usually associated with startups and small niche players.

Early in 1994, Ginn sent a down-to-earth message to employees entitled "How We Do Work Around Here." He highlighted the organization's core values: customer focus, innovation, respect for people, careful cost management, commitment to total quality, and high performance. Customers were the key to success. "We think about them as individuals and serve them one at a time," Ginn wrote.[17] He stressed the importance of showing respect for people at all levels of the business, from new hires to customers to the community. Yet "How We Do Work Around Here" was not prescriptive. It was an invitation to employees to create a new culture that would give them more freedom in their work lives and a sense of pride about their jobs. Ginn was reaching out.[18]

When he did, he tried to provide the leadership that would keep these values from becoming tiresome platitudes. Over the years, he had encountered CEOs and other executives who didn't take their cultural responsibilities seriously. Those who did were aware that their own behavior was magnified within the organization a hundred times as stories flowed through the firm. In brief, the iceman had to melt, and he did. On the first morning he arrived for work at the new offices, Ginn stepped out of the elevator and was greeted by a newly hired receptionist, Monica Dotson. She stood up from her chair, offered a full-armed wave over her station, and cried "Goood morning, Mr. Ginn!" Ginn smiled. To the surprise of onlookers, he offered back the same full-armed wave and called out, "Goood morning, Monica!" Then he went to work.[19] Over the next several months, he continued to work hard at changing his personal style of leadership. No longer formal and distant, he added a touch of humor to his interactions with employees. In meetings he listened more, encouraging greater candor. It was not always easy to give up the command-and-control style that had long been a central part of his career. In some ways, he had to change more than any other executive on the team. He was, after all, a product of AT&T's elite corps – the Long Lines Division – and thus a quintessential and successful Bell-head. Cox had always resisted the AT&T way, like many Pacific Telephone executives, and by 1994 he had been in the competitive world of wireless for nearly seven years. Sarin came to the business after divestiture with a business-school perspective on management and, despite his years with Pacific Bell, held to his enthusiasm for a competitive world. Neels, who was president of the international business, had developed his career in an international business milieu. So it was Ginn who had to push himself to take greater risks and trust employees to use their own judgment. He was not always successful.

One day, as he and his staff were discussing the company's values, someone suggested that the company should recognize employees' needs for a balanced life. For a moment, Ginn froze back into a Bell System posture. He was reluctant to open what he thought might become a can of worms for supervisors who were trying to hold their employees to their responsibilities. "I don't want someone coming back to me saying, 'You say a balanced life is important, but my supervisor just told me I can't get two hours off tomorrow to go see my kid in a play.'"

Amy Damianakes, his director of corporate communications, intervened: "Okay, this is a tricky one," she said. "But when something is sensitive, you're better off to get out there and define what you mean by it, instead of letting people think they know what you mean."[20] Ginn listened and met the issue head on. He acknowledged the employees' needs for flexibility but articulated a shared responsibility to ensure that the needs of the customer and the business were met. One way to do that was cross-training, so that employees could cover for one another as a team. That, he said, would be the AirTouch way to provide personal flexibility.

Given the complex web of partnerships and joint ventures that characterized the wireless industry and the company's business, Ginn faced an enormous challenge inducing employees to identify with the firm. After all, a given employee might move from the company's cellular operations in California to the Omnitel partnership in Italy – and then come back to the Cellular One joint venture with McCaw. This kind of mobility made the emphasis on a shared culture, employee communications, and programs like the two-by-four stock plan especially important.[21]

Decentralization was the key to operations. The market structure of cellular in the early 1990s focused the competitors' attentions locally, whether in San Diego or Dusseldorf. Ginn liked this. He believed that local people knew best how to respond to local trends, local politics, and local competition. He and his executives, led by Sarin, centralized the financial management of the business and other functional activities that generated economies of scale, but they did not try to tell managers how to build their businesses. Ginn relied on Lee Cox to motivate his managers to take charge of their own destinies, and Cox was tremendously successful.

These cultural and operational innovations were especially effective when those managers and their employees combined them with the engineering tradition of the Bell System, which remained strong at AirTouch. Bell capabilities helped them through a critical transition in technology. Still respectful of the management training he had received in the Bell System, Ginn continued in the Bell style to devote a great deal of care to the careers of junior managers.

Despite these accomplishments, AirTouch was still on the steep part of its learning curve. Compared to McCaw in the United States or Vodafone in the United Kingdom, Ginn's new company was still a cautious and not very efficient innovator. For example, an internal study in the second quarter of 1993 showed that, while McCaw averaged nearly 60-percent incremental cash-flow margins in 1991 and 1992, PacTel managed only 35 percent. PacTel had not broken out of the pack of Baby Bells. It ranked sixth, for example, among its siblings in penetration in its markets, and sixth out of seven in subscriber growth in the same period. Only one PacTel market (Atlanta) beat the industry average. One key measure boosted PacTel's success: its average revenue per customer (ARPU), where PacTel ranked ahead of all its siblings but behind McCaw and LIN Broadcasting Corporation. Here again, Los Angeles made an enormous difference, with average revenue per customer almost 25 percent higher than the industry average.[22] The legacies of the Los Angeles Olympic Games were still providing the enterprise considerable momentum.

Ginn and the other AirTouch executives were determined to catch and then distance themselves from the rest of the Baby Bells. In arguing for the spin-off, Ginn had maintained that a stand-alone wireless company would increase management's focus and efficiency. At the other Baby Bells, decisions were often delayed as one part of the business vied with another over turf or prerogatives. Ginn hoped that AirTouch would be able to abandon that part of the Bell way and become, in the process, a more nimble innovator.[23]

They didn't have long to make their transition and improve the efficiency of AirTouch operations. The world of telecommunications was, they knew, experiencing a radical transformation as the pace of deregulation and liberalization quickened both at home and abroad. Entrepreneurship in this setting would continue to place unusual demands on firms and individuals to be flexible and opportunistic. During the First Industrial Revolution of the eighteenth and early nineteenth centuries, change had been measured in decades. During the Second Industrial Revolution it came in years. But now, in the Information Age, decisive changes were taking place every month. In this new business universe, the best-informed individuals could barely keep up with the pace of change, and governments seldom could. AirTouch needed to be a creative organization from the top to the bottom if it was to master the unusual political economy of wireless in the 1990s.

SEVENTEEN

Competition from New Quarters

From his new office at One California Street, Sam Ginn could look out over the foot of Market Street, past the hulking structure of the Hyatt Regency, topped by a revolving restaurant. He could see the elegant clock tower of the 1898 Ferry Building rising against the backdrop of Treasure Island and the Bay Bridge. Cargo ships came and went from all over the world, plying the blue-green waters of the bay as sailboats and ferries dodged their wakes. These ships, their decks loaded with containers, seemed strangely anachronistic. They represented an earlier economy, one that the Information Age appeared to be rapidly leaving behind.

Wireless was already on the cusp of another major transition in 1994. Demand for cellular and paging services was growing fast as new groups of customers entered the industry with different expectations. As the cost of owning and using a cellular phone fell to about half what it was in 1983, soccer moms and teen-agers were joining the cellular ranks previously dominated by real-estate agents, doctors, and business executives. Many of these new consumers bought their first phones for emergencies, but as usage prices dropped, more people were making calls on their cell phones simply because it was convenient.

There were still uncertainties ahead. Just as car radios had been blamed for distracting drivers and causing accidents in the 1930s, critics charged that cell phones increased the hazards of the roadway.[1] Media stories raised fears that microwave transmission and electromagnetic radiation so close to the brain might cause cancer, and it was possible to imagine that some new discovery might sink the industry. But two major studies published in 1992, one by the U.S. Food and Drug Administration and the other by the IEEE, failed to find any evidence of risk. A report issued by the U.K.

National Radiological Protection Board concurred.[2] Although health fears may have dissuaded some people from using their cell phones, clearly the majority of consumers were reassured.

Other changes in the technology and pricing of cellular were also driving usage while changing the social perception of the cell phone. Here the primary factor was the rapid decline in the cost of portables. The "brick" that Sam Ginn had used at the 1984 Olympics had cost about $1,000 more to buy than an installed car phone. By 1993, however, the cost of a portable phone was nearly the same as a car phone – each could be purchased for only a couple hundred dollars.[3] The proliferation of portables made the cell phone more attractive to a wider variety of users. Working women, for example, increasingly bought cell phones not only for safety reasons but also to enable "remote mothering" of their children.[4]

The emerging anytime–anywhere world seemed to create an imaginary bubble of private space in a public environment. But this was obviously an illusion. Bystanders resented the intrusion of the private world into public space. As one sociologist noted, bystanders were treated as nonpersons by the phone user and were suddenly forced to pretend they weren't listening to the conversation. Patrons at theaters, concerts, and restaurants resented the disturbance created by the ringing or chirping of cellular portables. Miss Manners published guidelines to help cell-phone users negotiate the social parameters of the new technology.[5]

Despite these social tensions, AirTouch management knew that a growing number of customers clearly enjoyed the freedom that wireless gave them and that the service's increasing popularity gave the firm an unusual opportunity for expansion – if they could best their competition. The competition was growing and changing in ways that posed a variety of threats. AT&T's acquisition of McCaw promised to shift the geographic basis of competition from scattered regional metropolitan areas to a single national market. Meanwhile, government regulators were intent on expanding the number of competitors in each market.

Both the FCC and the state regulatory commissions had refrained from heavily regulating cellular prices, hoping that the duopoly structure they had created would stimulate competition. Besides, they reasoned, cellular was a nonessential service and therefore not appropriate for regulation. But in the late 1980s, at both the state and federal levels, consumer complaints about the high cost of cellular had prompted regulatory review. As usual, the California commission had jumped first off the mark. The CPUC had launched an investigation in the fall of 1988 to determine whether the regulatory framework should be changed. Their report, issued in June 1990, concluded that prices for cellular service were generally based on the market value of the service rather than cost or competitive pricing. In some cases, the report asserted, companies were earning returns on investment of over 50 percent. The Commission also expressed concern that cellular companies were purposefully repressing demand in order to avoid investing in additional network capacity.[6]

The Commission considered imposing rate-of-return regulation, but the majority of the commissioners rejected this proposal. In the wake of the Reagan revolution, regulators across the country were backpedaling from this kind of rigid control. In a move that was to become all too characteristic of the agency, the Commission called for greater competition but took only limited steps to remove the regulatory red tape that impeded the market. Instead, it added to the bureaucratic jumble a rule requiring cellular companies to (a) expand their systems as rapidly as possible and (b) lower prices to fill their capacity.[7] This regulatory innovation left everyone, including consumer advocates and some members of the legislature, unhappy.

When the federal government entered the fray, cellular prices once again grabbed headlines in California. A Government Accounting Office (GAO) study of the cellular industry confirmed many of the CPUC's conclusions. Given the duopoly structure, the regulatory barriers to new entrants, the lack of substitutes for cellular service, and the minimal differences in quality, the GAO said cellular prices were seldom priced competitively. Despite the head start of the wireline operators in the business, penetration rates and overall subscriber volumes for wireline and nonwireline companies around the country were roughly equivalent. In about two thirds of the markets, prices and packages offered by the two service providers were similar and had remained basically the same from 1985 to 1991.[8] The curious structure of the industry, which made companies like McCaw and PacTel partners in one market (San Francisco) and competitors in another (Los Angeles), tended to temper competitive behavior.[9] Overall, the government concluded, the duopoly structure had not created a good situation for consumers.[10]

The cellular industry, of course, rejected the idea that companies were earning excessive profits. High capital costs to build cellular networks, combined with a constant need to expand facilities to keep up with demand, meant that most cellular network operators in 1992 were not yet cash-flow positive. McCaw, for example, had yet to show a dollar of profit. In addition, most cellular providers faced a problem of limited capacity in dense urban areas. Lowering prices would increase demand, but their systems could not handle many more additional customers without significant new investments.[11] The industry was continuing to struggle to raise the capital needed to fund expansion.

Instead of additional regulation, both the GAO and the FCC looked to competition for solutions to the problems they perceived. They hoped that the imminent development of a new generation of personal communications services, PCS, would bring new entrants into all of the wireless markets. PCS was developed as a concept rather than a specific technology in the late 1980s. According to the engineers, PCS could utilize a patchwork of thousands of microcells with low-power transceivers mounted on buildings, phone poles, or even inside subway stations. With these transceivers close at hand, a new generation of handsets wouldn't need the same power as conventional cellular equipment. Batteries could be smaller and would operate longer. Phones would fold up, fit into a shirt pocket, and be cheap enough for millions of people to afford. With digital technology, PCS phones would allow people to use a single phone number that would follow them wherever they went.

Demand estimates for PCS ranged wildly, with some people projecting 60 million users within ten years. No wonder *The Economist* described PCS as the wireless industry's "Holy Grail."[12] Reed Hundt, chairman of the FCC, had a vision of economic democratization. With PCS, he said, "Wireless phones will no longer be status symbols of the rich and famous."[13] But as the concept developed in the early 1990s, the idea of PCS as a service distinct from cellular began to blur. Many began to see it as simply the next generation of cellular technology. Licensing new companies to provide PCS simply constituted "wireless deregulation." As one PacTel study concluded in 1993, PCS would almost certainly bring intense competition to the wireless industry.

Indeed, the prospect of PCS excited a host of potential new competitors. Businesses and entrepreneurs who had missed out on the first round of cellular licenses, including many cable companies, asked the FCC for special treatment in this next wireless generation.[14] Long-distance providers Sprint and MCI, which had sold their first-generation cellular properties, now had second thoughts about the industry and wanted back in through PCS. Competition also sprang up on another front: the FCC had decided to allow a new breed of dispatch service providers to offer cellular-like service on spectrum that had been reserved for specialized mobile radio (SMR). In this new sector, the principal competitor was Fleet Call (soon to be reborn as Nextel). Founded in 1987 by a former FCC lawyer and a cellular executive, Fleet Call had quietly gone around the country buying up rights to taxi and truck radio dispatch frequencies. Then the company asked the FCC for permission to offer phone service on these channels. Though cellular providers cried foul, the FCC approved this request in 1991, and the firm, now Nextel, promised that it would take the cellular industry by storm.

During the AirTouch road show, executives who were trying to raise capital had felt they were shadowed by the forthcoming IPO of Nextel, which claimed that it would bring lower prices to the industry and take away significant market share from the incumbents. Analysts, who were already concerned about the threat of PCS, found the Nextel presentation particularly troubling. Ginn, Cox, and Sarin tried to convince these skeptics that they would be able to provide a higher-quality product at a comparable price and retain their existing customers. But they would rather have been on more positive ground.

At first AirTouch and other cellular leaders weren't really too concerned. Since Nextel used traditional analog technology, it didn't seem to represent a great threat to cellular. The licenses Nextel held typically included only 10–15 MHz of spectrum (compared to the 25 MHz allocated to each cellular licensee), so Nextel would inevitably face capacity problems. Moreover, Nextel's spectrum was not contiguous, and subscribers might have to wait ten to twenty seconds after they turned on their phones for the equipment to find an open channel.[15] Ginn and his chief technology officer, Craig Farrill, had tried to make these points to analysts during the IPO road show, but in city after city, the analysts didn't buy their pitch. They pointed out that, by converting to digital, Nextel could solve its capacity problems. Given its potential for nationwide coverage, Nextel threatened to become a major competitor. With the emergence of Nextel and the combination of AT&T and McCaw, the race to assemble national wireless footprints accelerated.

Now deeply concerned about the new strategy, AirTouch began to look around for potential acquisitions that would enable the company to build a national franchise. The management team kept looking to Washington, DC. New spectrum and new licenses might give them the opportunity to expand AirTouch's coverage dramatically – but that, of course, depended on the FCC's next move.

*

The controversy over PCS licensing was protracted and rancorous. The FCC had more questions than answers. What spectrum should be allocated for this new service? What spectrum users would have to be displaced to make room for PCS providers? How should licenses be allocated and on what geographic basis? Should the government consider issuing national licenses as governments in Europe had done? Or should it stick with the metropolitan trading area concept used in the first wireless generation? Who should be allowed to compete for these licenses? Should existing cellular companies be included or excluded? Should the FCC mandate a technology standard, as it had done with AMPS and the first generation of cellular in the United States? Or should it let competition work out the technological standards while also fostering innovation and lower prices? Billions of dollars were at stake, as well as the future of the wireless industry.

If the FCC was gun-shy, it had good cause. Tremendous criticism had been heaped on the agency for the way it had awarded first-generation licenses. As one writer suggested, the process was not much different from a state lottery. "They even use one of those ping-pong ball machines to pick the lucky number." But instead of raising money for the government, the spectrum lottery cost the government about $20 billion. For the players in spectrum lottery (called "Phone Lotto" by some), the odds of winning had been good. Where a typical state lottery paid out about 50 cents on the dollar, the spectrum lottery netted winners about $20 for every dollar they invested in applications.[16] Stung by these criticisms, the FCC searched desperately for a better way to award PCS licenses.

Auctions offered a possible solution. Nobel laureate economist Ronald Coase and several other academics had suggested as early as 1959 that the government should auction radio spectrum. Auctions, Coase argued, would provide just compensation to taxpayers for the use of a common resource. They would also rationalize and streamline the award process, taking politics out of the equation. Speculators would no longer reap quick rewards on a free public good; instead, the taxpayer would benefit from the government's capture of the value of spectrum. By eliminating the long period of uncertainty that went with bureaucratic allocation, auctions would provide greater certainty and encourage investment.[17]

Broadcasters, of course, didn't embrace this kind of competition. They preferred a government giveaway and argued that selling spectrum to the highest bidder would undermine the important civic responsibility of broadcasters to maintain the airwaves as a public trust. Their political pressure on Congress blocked the FCC's efforts to auction radio spectrum.

In the mid-1980s, however, the FCC returned to the idea. Chairman Mark Fowler pitched the concept to Congress as a way to allocate the next generation of cellular licenses. Some opponents countered that an auction would ensure that only large, deep-pocketed companies would win. These were not the companies, they said, that would be innovators in this new and exciting industry. Consumers would suffer. Auctions would also raise the capital requirements of the industry, resulting in higher prices. Lobbyists in the broadcasting and telephone industry again managed to defeat legislative efforts to sanction spectrum auctions. By the early 1990s, the FCC and Congress were deadlocked over the issue. The FCC refused to reallocate spectrum for PCS until Congress gave it auction authority. Congress refused. Finally, the Clinton administration's efforts to reduce the federal deficit broke the deadlock.

The Budget Reconciliation Act of 1993 transformed regulation of the cellular industry and set the stage for broader telecommunications reform down the road. Congress explicitly barred the states from regulating either entry or prices in wireless. There would be no return to rate-of-return in this industry. The legislature also gave the FCC authority to control interconnection between wireless companies and local telephone firms.[18] At the same time, Congress approved the PCS auction (though still forbidding its use for new radio or TV licenses) and ordered the FCC to shift at least 200 MHz of federal bandwidth into the private sector.[19]

In designing the auction, the FCC faced a formidable task. No one had auctioned radio spectrum successfully. New Zealand had tried in 1990 but had met with sharp criticism when the process failed to return the estimated value of the licenses. The FCC could not just auction the licenses one by one and be fair to bidders, because values were interdependent. A license for suburban San Jose was worth more if you obtained the license for San Francisco as well. To develop an adequate structure for the auction, the FCC turned to game theorists at Stanford and the University of Texas.[20]

Hungarian-born math wizard John von Neumann, one of the fathers of modern computers and the nuclear bomb, had developed game theory in the 1940s.[21] Game theorists sought to predict the interactions of multiple players in competition ranging from something as innocent as a bridge game to something as deadly as nuclear war. At first considered an arcane field of study, game theory became increasingly relevant to military and business strategists in the 1970s and 1980s as computers made it possible to consider the effects of almost limitless permutations. Theory and reality meshed in 1994 as the FCC adopted a game-theory strategy for the auctions, and in the same year three game theorists shared the Nobel Prize in economics.[22]

The advisors designed an auction with simultaneous ascending rounds that allowed bidders to compete for any or all the PCS licenses at once, thus recognizing the synergistic values in owning specific combinations.[23] It was an open bidding process, so participants could see how their competitors' strategies were evolving after each round and then adjust. According to the professors, this kind of auction would help bidders arrive at the right prices, award licenses to companies that would use them efficiently, and produce the most revenue for the government.[24] The government added rules designed to prevent companies from "sandbagging" (a term well-known to poker players) by bidding high on the first pass and then withdrawing a winning bid. The FCC also discouraged speculators by prohibiting "unjust enrichment" in the sale of licenses acquired at auction and requiring companies to make substantial up-front deposits. Altogether, these deposits, submitted by thirty bidders, totaled more than half a billion dollars. This was a high-stakes game.

The FCC tested the concept in the summer of 1994 when it auctioned spectrum for narrowband PCS pager services. The experiment, which from the FCC's perspective was a remarkable success, netted the government $617 million. As the end of the year approached, the agency prepared for the main event, the sale of 99 broadband PCS licenses covering 51 metropolitan areas. Some observers predicted that this would be the largest public auction ever held and might net the government anywhere from $10 to $20 billion. But that, of course, would depend on the strategies and resources of the bidders.

*

As the rules for the auction became clear, AirTouch executives grew worried. A bidding frenzy might make PCS licenses so expensive that it would take AirTouch years to recoup the capital invested. Already, long-distance companies like MCI and Sprint had announced that they planned to pursue PCS aggressively. Cable companies had been knocking at the FCC's door, and Ginn was sure that all of the Baby Bells would be in the running, along with AT&T and a host of smaller companies. With widespread reports in the media that PCS would attract tens of millions of customers within a matter of years, it was no wonder that some people feared a bidding circus. "No one has a prayer of making money," one analyst said, "if the bids go too high."[25]

No one could be certain how much it would cost to build PCS networks, and this uncertainty shadowed the auction. FCC Chairman Hundt predicted "the greatest one-time private sector investment in any single industry in the nation's peacetime history" – an estimated $50 billion.[26] That frightening figure didn't even include a major hidden cost of PCS. To free spectrum for the new service, the FCC planned to bump utilities, railroads, and public service agencies to a different part of the radio band, but it decided that PCS operators should pay to move them. No one knew what this would cost. All of these issues gave the AirTouch auction team sleepless nights. On the one hand, an expensive auction would be good news for AirTouch because it would give the cellular operations a cost advantage over PCS services. On the other hand, it would limit the company's ability to expand its footprint and compete with AT&T and Nextel. Although AT&T would enter the auction to pick up licenses to fill in coverage, with McCaw's network it already had a commanding lead in the race for national coverage. AT&T disdained other partners. Clearly, they were the company to watch.

Ginn and his colleagues tried to simplify their job by forging alliances with one or two other major cellular operators. Then, they reasoned, they would only need to fill in gaps at the auction. With a strong presence in the West, the Midwest, and the South, AirTouch was particularly interested in the East Coast, so they turned to Nynex as a potential partner. The two companies were on the verge of an alliance in the spring of 1994 when a surprise announcement from MCI caught Ginn and Nynex CEO Ivan Seidenberg off guard.[27] Disappointed by the FCC's refusal to issue a nationwide license for PCS, MCI had decided to avoid the auction altogether. Instead, the company decided to fuel Nextel's assault on the wireless industry with $1.3 billion. Already dogged for months by Wall Street's fixation with this upstart company, Ginn and Seidenberg hit the brakes.

When the industry calculated the significance of the MCI–Nextel deal, a flurry of corporate courtships followed. Nynex rejected AirTouch and fashioned an agreement with Bell Atlantic. The two companies promised to combine their cellular operations and bid together for PCS licenses. Not to be outmaneuvered, AirTouch struck a deal with U.S. West, agreeing to merge their domestic cellular assets. This didn't give AirTouch the East Coast presence it wanted, but it gave the company licenses in nine of the top twenty U.S. markets, covering more than 53 million people. The combined entity would be a formidable player in the industry, and based on the relative value of its POPs, AirTouch received 70-percent ownership. In addition to its domestic cellular interests, U.S. West brought to the table valuable experience from the U.K.'s PCN market, where it held a 50-percent stake in Mercury's newly launched One 2 One network.[28] That experience would help AirTouch and U.S. West develop a successful joint venture to bid for PCS licenses.[29]

As the FCC's October 28 deadline for registration approached, the mating dance got faster and more complicated. *Fortune* compared the telecom companies to "gerbils, coupling and uncoupling *en plaine vue.*"[30] In August, MCI reportedly lost confidence in Motorola's technology and Nextel's ability to deliver high-quality voice services over its dispatch network and spurned Nextel at the altar.[31] The press reported in September that Bell Atlantic and Nynex were talking to Sprint. A deal between the three companies would bring together 3 million wireless users and surpass AT&T–McCaw as the nation's largest cellular enterprise.[32] But these talks also collapsed.

In the midst of this flurry, Arun Sarin planned to celebrate his 40th birthday at the Auberge du Soleil in the Napa Valley. Close to sixty people had arrived for the celebration when Sarin got a call on his cell phone. The AirTouch corporate development team in New York had heard that Bell Atlantic and Nynex were very close to a deal with MCI. "You've got to come out here now to try to stop this thing right away," his staff told him. In the middle of the festivities, Sarin and Lee Cox compared notes. If MCI forged an alliance with Bell Atlantic and Nynex, then AirTouch and U.S. West would be forced to align with Sprint. These combinations would give the long-distance companies a central, dominant position in the wireless consolidation movement. Cox and Sarin agreed that they had to try to break up the deal. Leaving his guests, Sarin headed for the airport and a flight to New York.

The next day, he met with executives from Bell Atlantic and Nynex. With MCI's representatives in the room, he pressed the strategic issues, arguing that the natural alliance was to put the two East Coast Baby Bells together with AirTouch and U.S. West. That would create a formidable national footprint. The MCI representatives were, of course, irritated by Sarin's interference – especially after their deal fell through.[33]

Sprint, in the meantime, had turned its attention to a consortium of cable companies that had come together in December 1993 to bid for PCS licenses. Bill Esrey, Sprint's CEO, seemed committed to the convergence of the fixed wire and wireless businesses and was certainly determined to have a nationwide system. Hoping to bypass the Baby Bells in the local exchange business, he and the cable companies agreed to join together to bid more than $3 billion for PCS.[34]

Determined not to be left out, AirTouch's executives continued negotiating to create a blockbuster deal. In October, Ginn and the CEOs of U.S. West, Bell Atlantic, and Nynex announced that they had reached an understanding and were forming a joint venture. Their alliance would stitch together a customer base of 4 million people, with the possibility of reaching 100 million.[35] Each of the partners had substantial experience in deploying digital cellular networks as well as significant ability to raise capital. If the joint venture (dubbed PCS PrimeCo) hung together, it would change the entire landscape of wireless in the United States.

Since the spin-off, Bell Atlantic had approached AirTouch on a number of occasions regarding a deal. Ginn had rebuffed each of these inquiries, but as the importance of a national network became apparent to everyone in the industry, AirTouch and Bell Atlantic began to edge closer. For a time, they eyed each other like wary wrestlers at the beginning of a match. But neither side wanted to be left behind in this national turf struggle and, as part of the PrimeCo contract, Bell Atlantic and AirTouch agreed to a truce. Neither would make a play for the other. But they both understood that eventually a single national wireless entity would emerge from this partnership. The big question was, who would be in control?

The partners allowed AirTouch, out of deference to its experience with digital, to nominate PrimeCo's CEO. With so much riding on the outcome, Ginn turned to one of his own most experienced and trusted executives. George Schmitt was still living in Dusseldorf in 1994, but he was rarely home. After the successful launch of the D2 system in Germany, he had become executive vice-president for international operations. In this position, he was responsible for the development of more digital wireless networks than any business executive in the world. He divided his time between Europe and Asia, with occasional trips to California. Lee Cox had told Schmitt about the PrimeCo deal and asked if he would be interested in being CEO. Schmitt said "No." He liked his current job, and when they were able to stay at home, he and his wife were comfortable in Germany.

Ginn persisted. As the combinations solidified, he summoned Schmitt and the other members of his policy group to a meeting at the historic Claremont Hotel in Oakland. During a break, he asked Schmitt to talk to him privately. They met in Ginn's room, which offered a view over the hotel's gardens and tennis courts, the eucalyptus and palm trees, past the white bell tower and beyond to the urbanized neighborhoods of Oakland and Berkeley. Ignoring the scenery, Ginn quickly leaned on Schmitt to accept the new position.[36] No one in the cellular industry, he said, had Schmitt's experience or ability. Schmitt had deployed more digital networks, knew more about handling suppliers, and had demonstrated a rare ability to keep a project on schedule. With him as CEO of PrimeCo, AirTouch's position would be strong. Despite the praise, Schmitt was unconvinced. Invoking his usual four-letter vocabulary to clarify his case, he suggested that Ginn look to someone else. But Sam's mind was firm, and he made it clear that he was not really offering a choice. Reluctantly, Schmitt agreed to be Ginn's candidate. After passing muster with the other partners, Schmitt took over as PrimeCo's CEO in time for the December auction.[37]

In Washington, DC, the telecom mating dance was causing some distress. Consolidation was an unanticipated result of the auction process. Neither the professors nor the regulators had given this consequence any thought, in part because they analyzed the process in neoclassical, static terms. Dynamic, structural changes were left out of their assumptions. But they were not left out of the assumptions of the executives who were going to invest billions of dollars to buy the right to test an as-yet unproven technology. The executives wanted to cut their risks and improve their odds of success. They narrowed the field of potential bidders and perhaps limited the demand for specific markets. The major companies now only needed to fill in gaps. This raised some eyebrows at the Department of Justice, which of course launched an investigation into these "mega-alliances." Few expected the government's probe to change the outcome of the auction or even, as one analyst commented, to derail the "integration freight train."[38] Neither Ginn nor his new partners wanted any extended bouts with the DOJ, but they certainly didn't want to be left at the station when the "freight train" left.

As the December 5th start of the auction approached, the strategies of the major players were clear. Three bidders were pursuing nationwide networks: Sprint, AT&T, and PCS PrimeCo. Of these, Sprint and its cable partners laid down the biggest deposit, $118 million, which allowed them to bid for up to 197 million POPs.[39] Other players targeted regional markets. Pacific Telesis Group, which was now headed by Phil Quigley, was looking to get back into the wireless business. It left no doubt that it would spend whatever it needed to win Los Angeles and San Francisco, where its existing infrastructure gave it a cost advantage over all other competitors.[40] Several other telecoms decided they would not spend anything on the auction. In the end, MCI stayed out. The other Baby Bells – including Bell-South, Southwestern Bell, and Ameritech – pursued only markets within their existing regions.

Among all the potential bidders, Craig McCaw was the most enigmatic. Having sold his company to AT&T, McCaw had watched the corporate coupling with detached amusement. But then, realizing that competition might be substantially reduced and values improved, he decided to get involved. He formed his own bidding entity, ALAACR Communications, with the intention of bidding in 22 regions. Some people at AirTouch reasoned that McCaw would support AT&T's interests.[41] But as the auction began, he remained the most inscrutable of all the players.

The bidding began cautiously in December. Despite the transparent character of the process, bidders searched for ways to hide their cards. To remain eligible for all the POPs they eventually hoped to win, they had to bid on that many POPs in each round. This led to an elaborate tactic of looking for places to hide bids.[42] Thus, to hold down the bid price in Chicago in the early rounds, PrimeCo pretended not to be interested. In order to maintain its eligibility, it entered bids for markets like St. Louis, Little Rock, and Honolulu.[43] The FCC was concerned about the possibility of collusion. But while the bidders signaled like they were playing bridge, there was little evidence that it did any good. In California, despite the posturing of Pacific Telesis, Craig McCaw raised the stakes dramatically when he jumped the bidding for Los Angeles from $183 to $300 million in one round. "I could not conscience that they would just steal these markets," he said. "I mean, the FCC would never forgive me for not bidding, and I felt it was my duty to go in and blast them ... I didn't torment them too long. They were such good markets, I'd have been willing to pay far more."[44]

As it became evident that Chicago would be a major battleground, Schmitt warned bidders away. "You mess with me in Chicago," he told the *New York Times,* "you pay."[45] But no one seemed intimidated by these declarations. To Schmitt's dismay, the Sprint consortium, AT&T, and PrimeCo bid so aggressively for Chicago's two licenses that they drove the bid price to the highest level in the nation.[46] If there was collusion, it was a well-concealed failure.

McCaw continued to shake things up. As the bidding escalated, however, he saw his opportunities diminishing. The prices were going too high. Other bidders began to drop out. Then McCaw finally withdrew.[47] As one commentator put it, "He was an opportunistic bidder, who in the end did not find any opportunities."[48] Despite his lack of success, McCaw made a great deal of money for the U.S. government. Single-handedly, he raised prices in several markets by somewhere between $500 and $800 million. Rather than helping AT&T, McCaw seemed to be guided by the central assumption that had held through his business career: he saw the inherent value of owning spectrum. He was looking for a good deal wherever he could find it. Sometimes frustrated and sometimes amused by McCaw's role, Ginn told *Forbes* that "Craig McCaw thought he was playing Nintendo. He was having the time of his life."[49]

After nearly three months and 112 rounds, the auction finally ended in March 1995. Sprint, AT&T, and PCS PrimeCo walked away with the biggest wins. Sprint and its partners spent $2.1 billion to acquire 29 licenses covering 145 million POPs.[50] At a cost of $1.68 billion, AT&T picked up licenses in 21 markets with 107 million POPs. PCS PrimeCo gathered 11 licenses covering 57 million POPs, spending a total of more than $1.1 billion for Chicago, Dallas, Tampa, Miami, New Orleans, Houston, Milwaukee, Richmond, San Antonio, Jacksonville, and Honolulu. Some people wondered whether PrimeCo had paid too much because it spent more than $19 per potential customer for its licenses, about $4 more than the auction average. That figure became even more frightening when you realized that the consortium would have to spend another $1.4 billion to build the networks for these markets.[51]

From a short-term policy perspective, the broadband PCS auction was successful. It was the largest public auction ever held. Although it took three months to complete, that was less time than the FCC had taken to dole out the first thirty nonwireline licenses at the dawn of the cellular era. The bids netted taxpayers nearly $7.7 billion.[52] The auction was indeed a neoclassical economist's dream. Marginal costs and prices appeared now to be hovering together and, it could be assumed, the goal of allocative efficiency was achieved.

From a long-term, dynamic perspective, the results were more difficult to evaluate. Instead of pushing prices down toward costs, the auction had pushed marginal costs up toward prices. It was not likely that consumers would benefit from that change. Nor, for that matter, would the industry or the economy benefit from the fact that the auction marked the end of the intensely entrepreneurial era of wireless development, an era that had spawned new competitors like AirTouch in the United States and Vodafone in the United Kingdom. From the auction on, consolidations and structural change would be the central theme of wireless evolution, first in the United States and then in the global economy. Global oligopoly would yield economies of scale, but it would, as always, distress the same sort of neoclassical analysts who had been the staunchest advocates of the auction. Whether or not that would matter would be largely determined by the combines' performances – that is, by the quality of service they would provide, the prices they would charge, and the innovations they would introduce. In that sense, they held their future in their own hands.

The auction was certainly not a business executive's dream. All of the wireless firms had reason to be concerned. The PCS auctions had reshaped the competitive landscape. Suddenly, Sprint emerged as a formidable competitor. As FCC Chairman Reed Hundt explained: "The auctions just created the single largest wireless company in the world, and it's the cable television industry. This is the place where actual convergence between telephone, cable, and long distance is taking place."[53] Meanwhile, overseas, governments struggling with perennially tight budgets and the politics of license allocations watched the FCC's auctions with keen interest. Soon after the end of the auction, other governments abandoned beauty contests and followed the American lead. When this happened, AirTouch suddenly lost one of its greatest sources of competitive advantage abroad – the ability to win licenses on the strength of its track record and experience. From now on, in the race to blanket the globe with wireless, the lead increasingly went to those with the deepest pockets and strongest stomach for risk. As long as penetration rates continued to exceed analysts' projections, the risk-takers walked away heroes. But Ginn, Sarin, and the rest of the AirTouch team recognized that the pressure on management would increase rapidly. At home and abroad, the duopoly era – with profit margins that covered mistakes – had ended.

Ginn recognized that he had to reorient his leadership for this new era of wireless development. Jan Neels, who had brilliantly crafted successful partnerships in Europe and elsewhere, moved to a new job as president and CEO of the company's joint venture operation in Belgium. After more than two decades of globetrotting, this position brought him back to his homeland. Arun Sarin became president and CEO of AirTouch International. By 1995, he had demonstrated his strong operating skills. He could manage a tremendous workload, make decisions carefully and quickly, and recognize talent in others. He could work smoothly with managers and business partners in many different nations. The major challenge he faced was similar to the one he had confronted in the aftermath of the acquisition of CI: to begin systematizing AirTouch's various independent and often minority-owned companies in Europe and elsewhere. By consolidating operations – networks, billing, and customer care – the firm would generate global economies of scale.[54] It would need to be as efficient as possible, given the problems that lay ahead in the large U.S. market.

EIGHTEEN

Battling Systems

PrimeCo and the other winners of PCS licenses in the United States quickly discovered how hard it would be to generate adequate returns from their investments. They were required to bear the cost of moving existing microwave users to other parts of the spectrum. The legislation gave these incumbents a great deal of leverage, and they were – PrimeCo and other would-be PCS operators complained – making extortionate demands. There were other serious problems as well. Because PCS used microcells, system builders needed many more locations for receivers than traditional cellular operators. They had to negotiate with more land and building owners for cell sites. Even when they could cut deals, the operators increasingly ran into problems with communities opposed to unsightly antennas. To get PCS running, the federal government had to step in and require the Post Office to provide the operators with space for their antennae. Still, costs kept mounting, and PrimeCo's executives became concerned that the increases would soon eviscerate the savings PCS was supposed to offer to consumers. The margins were shrinking at a dangerous rate.

While confronting these issues, they had to deal with a difficult technological situation: they had to pick one of the emerging digital transmission technologies in order to move forward with PCS. The FCC had punted on the question of a standard technology and decided to let the industry somehow settle the matter. For the domestic wireless industry and Air-Touch's future, the consequences of this choice would be enormous. Standardization abroad, especially in Europe, fueled the pace of innovation. But in the United States, the industry spent the next several years arguing over the merits of competing technologies and, in a throwback to the early days of telephone history, developing different systems that wouldn't work together.[1] Every month that this situation existed, the United States slipped further behind Europe in the deployment of new digital systems and the development of new products.

The transition to digital was being driven by increases in demand. Research on the development of the digital technology that would accommodate more users had started almost as soon as the first AMPS systems began operating in the United States in the mid-1980s. Traditional analog service divided the available radio spectrum into channels. Most carriers could theoretically handle up to 416 calls at one time within the 25 Mhz of spectrum allocated to them by the government. But in many large urban areas, the demand for service had quickly exceeded this capacity. In Los Angeles, less than two years after PacTel turned on its system, it was almost impossible to complete a call on the first try during rush hour.

The advent of the truly portable phone had magnified the problem. The "brick" that Ginn had used at the Los Angeles Olympics in 1984 was so heavy and expensive that it didn't become popular right away. By the end of the decade, however, improvements in microchip capacity and battery technology had shrunk the size of the phone substantially. As portables became more popular, calling patterns began to change. Most cellular networks had been built to serve mobile customers in vehicles traveling along busy traffic corridors. Suddenly, customers were demanding service in shopping malls, downtown business centers, and other areas. In order to ensure quality coverage, PacTel had scrambled to build more base stations.[2]

The company continued to scramble and developed a microcell system that provided service inside buildings, canyons, and tunnels.[3] Then it converted to digital switching, which enhanced service quality and reliability.[4] But in Los Angeles, particularly, demand for service continued to exceed projections by 200 percent – year after year.[5] The FCC had responded by allocating limited additional spectrum for cellular.[6] But none of these efforts could solve the long-run problem. In urban areas of the United States, the demand for wireless was overwhelming the existing systems.

A full-blown digital technology offered a way out of this trap. PrimeCo and other companies could then divide each radio channel into time slots that allowed multiple conversations to use a single channel without interfering with one another. In essence, pieces of conversations sailed along the radio waves, accompanied by software codes that identified where they belonged. The base station and the mobile phone read the codes and put the pieces back together in a virtually seamless conversation.

But before the industry could switch to this new approach, called "time division," it had to settle on an interface technology that would allow the base station and the telephone to talk to each other. In the United States, the Cellular Telecommunications Industry Association (CTIA) set out to establish the standard that was clearly needed. The organization developed parameters for the next generation of cellular in the mid-1980s, with a goal of introducing digital by 1991. As research and development progressed, a technology called time division multiple access (TDMA), developed by Sweden's LM Ericsson, emerged as the frontrunner. Both McCaw and BellSouth campaigned vigorously for its adoption. But there were some in the industry, particularly in San Francisco, who were reticent.

Still under the umbrella of Pacific Telesis in the early months of 1989, PacTel had talked to a fledgling company in San Diego that was taking a different approach. Qualcomm, which had been founded in 1985 to provide a satellite-based tracking and communications system for trucks and commercial vehicles, was interested in digital cellular technology as well. Its CEO, Irwin Jacobs, had been studying a concept called code division multiple access (CDMA). Adapted from systems used by the military to encrypt communications and resist enemy jamming, CDMA had been proposed for civilian use in the late 1940s but had never been developed. Jacobs suggested to PacTel's technologists that CDMA could be adapted for use in a cellular system.

Craig Farrill, PacTel's head of engineering, and William Lee, the company's top scientist, thought Jacobs was crazy. Twice they sent him away. But the second time, Jacobs asked, "What would it take for you guys to believe this was possible?" Farrill and Lee gave him a list of fourteen hoops to jump through. Weeks later, Jacobs was back with Qualcomm's results. "We turned our heads and said 'Wow, this may really be possible,'" Farrill remembers. PacTel agreed to become Qualcomm's first CDMA customer and to help sponsor development of the technology.

The CDMA technology had several selling points, the most important of which for PacTel was its capacity: it could handle a very high volume of traffic. Using what Qualcomm called "spread spectrum" technology, CDMA could support ten times the users of an analog system and three to five times the number of callers using a TDMA or GSM interface.[7] CDMA provided better voice quality and required less power, enabling customers to enjoy longer battery life while using smaller handsets. It also offered a smoother (or "soft") handoff as callers moved from one cell to another.

Fewer calls would be dropped. But as critics pointed out in 1989, most of these claims were untested.

Qualcomm promptly started the tests in November 1989 in San Diego. More than 300 people from throughout the industry showed up for a trial, which established that CDMA could accommodate ten times more customers on the same bandwidth as analog. The reaction from observers ranged from elation to disbelief. Tests in Manhattan the following February generated similar results.

But CDMA and Qualcomm were late to the party. When Qualcomm presented its system to the industry standards committee, the company got nowhere. Having just resolved a two-year debate over competing digital technologies by giving the nod to TDMA, the committee didn't want to start over again.[8] Manufacturers were starting to design and build equipment for this standard. The industry's ambitious goal of converting to digital by 1991 had already slipped to 1992 or 1993, and the committee wanted to avoid further delays. Given the momentum behind TDMA, industry watchers in 1991 predicted it would be the U.S. standard within eighteen months.[9]

Craig Farrill stubbornly refused to go along with the crowd. Once again, Los Angeles shaped the pattern of PacTel's decisionmaking. Most tests showed that TDMA could provide a tripling of capacity over the current analog system. In Los Angeles, PacTel estimated, that would meet the demand for only a couple of years, and then the company would once again be faced with a crisis. After Qualcomm demonstrated CDMA's viability on six PacTel cell sites in San Diego in the summer of 1991, Farrill was increasingly convinced that CDMA was the right choice.[10]

Without the backing of a government standards board, as in Europe, Qualcomm faced an uphill struggle to convince the cellular companies, one by one, to adopt its technology. As it poured resources into the development of CDMA, the company became stretched for cash, and PacTel became Qualcomm's white knight. The company bought Qualcomm stock and contributed about $2.2 million in development funds in exchange for warrants for 390,000 additional shares. Ginn even considered acquiring Qualcomm and bringing CDMA in-house. But PacTel's financial support for Qualcomm and CDMA had already elicited sharp criticism in the industry.[11]

Since the Baby Bells were still under the constraints of the MFJ, some opponents of CDMA were able to charge that PacTel's investments in Qualcomm violated the consent decree's ban on manufacturing. Others complained that PacTel would delay the rollout of digital, since product development for CDMA lagged TDMA by almost a year.[12] As the CTIA's executive director told reporters in January 1991, "It's time to shoot the engineers and go to market."[13]

Faced with the prospect of a protracted legal and regulatory battle, Ginn opted instead to seek allies for the company and the technology. Craig Farrill toured the Baby Bells and made the case for CDMA. He won over two converts: Nynex and Ameritech. Meanwhile Ericsson, the leading developer of TDMA, had launched an all-out campaign to discredit the new competitor. Craig Farrill faced intense pressure from his peers in the industry. "I was accused of being [Qualcomm CEO] Irwin Jacobs's patsy," he recalls.[14]

Even within PacTel there were differences of opinion about the technology. Opponents argued that CDMA had an Achilles heel, known in the industry as the "near–far" problem. Calls sent by customers with strong signals drowned out calls coming from weaker signals. Experts at Bell Labs and other research centers predicted in 1991 that the problem could not be solved.[15] At the same time, the business-school graduates in PacTel's corporate development group, backed by consultants at McKinsey, leaned on Lee Cox to join the majority, adopt TDMA, and, above all, get to market quickly. An internal struggle erupted between the MBAs and the engineers. Surprisingly, given the management culture of American business in the early 1990s, the engineers won.

Their victory was in part a testament to the residual power of the engineering culture that had permeated the Bell System for a century.[16] Also influential was Qualcomm's successful effort to solve the near–far problem. Using a series of feedback loops and smart handsets, Qualcomm transformed the "problem" into a competitive advantage. The new approach made the system a very low–power-consuming technology compared to GSM and TDMA. Batteries for CDMA phones could be even smaller and last longer – a key bonus for many customers.

Qualcomm's biggest break came when the government of South Korea and a consortium of Korean electronics manufacturers decided to bet on CDMA. Four of the country's leading electronics companies signed licensing agreements with Qualcomm allowing them to manufacture and sell CDMA phones and infrastructure worldwide. The Korean government also invited bids to license a second cellular operator in competition with the government-backed Korean Mobile Telecommunications Company (KMTC). The government "proposed" a CDMA standard for this new network. By betting early on CDMA, the government hoped to give

Korean manufacturers a chance to capture a significant share of the wireless infrastructure and handset market. AirTouch and its major Korean partner, the Pohang Iron and Steel Company (POSCO), eventually won this license.[17]

PacTel announced its decision to deploy CDMA at the CTIA show in New Orleans in February 1992. It was the first U.S. wireless carrier to commit to Qualcomm's technology, and it planned to spend an estimated $250 million to build its new digital networks in California and Georgia.[18] Ameritech and Nynex soon followed suit. As the momentum shifted, Qualcomm finally got a hearing before the industry standards committee. Solid support from the three Baby Bells and Qualcomm's technical accomplishments won the day. In 1993 the industry adopted a CDMA standard.

Unfortunately, the standards committee compromised in a manner better suited to political debates than technological progress. The committee continued to support its existing TDMA approvals.[19] This left the global wireless industry and the U.S. market in technological disarray. In network industries there is a substantial economic premium for standardization and interconnection. But now the United States was supporting at least three different digital systems and would remain separated from Europe by a special version of the digital divide.

Ginn was nevertheless so confident of CDMA that he told the industry in early 1995 that AirTouch's system would be available to customers in Los Angeles by mid-June. But unsolved problems with the base station software and with the communications between handsets and the base stations made that impossible.[20] CDMA was just not ready. Subsequent dates came and went, while McCaw, Bell Atlantic, and others began to roll out digital using TDMA. As CDMA lagged, the firm's hard-earned reputation for technical capability was placed at risk. According to one Stanford University professor, "They've got fundamental technical problems that they don't know how to solve."[21] He was right.

The debate over CDMA continued to cause serious dissension in AirTouch. George Schmitt, the CEO at PrimeCo, was an evangelist for Europe's digital standard, GSM. The operators and government ministries backing the GSM standard had elected him chairman of the GSM working group in 1993 with responsibility for leading the deployment of this technology throughout the world.[22] Schmitt was convinced that GSM would provide PrimeCo with its best and most cost-effective digital program, and he had little patience with CDMA's champions. He was convinced that they were overstating CDMA's potential. While Jacobs and Qualcomm continued to claim in 1996 that CDMA would deliver fifteen times the capacity of analog systems, Schmitt believed that the multiple was only "two or three."[23]

Schmitt pointed out that GSM equipment was available now. It offered a quicker time to market and thus lowered the carrying cost of capital invested in license acquisitions and network construction. Given the high volume of worldwide demand, GSM network equipment was cheaper and so handsets would be more affordable for customers.[24] Other PCS operators in the United States, including Pacific Bell Mobile Services, had already begun to deploy GSM. Nextel was testing GSM in Los Angeles, San Francisco, Seattle, and Denver.[25]

Schmitt and Craig Farrill, who was by now president of the worldwide CDMA Development Group, debated the issue with Lee Cox. Farrill remained convinced of CDMA's technical superiority, but he was on less solid ground when it came to the economics of the system. Farrill argued that CDMA would require substantially fewer base stations overall, only one quarter of what GSM would need. Schmitt countered that CDMA base stations were expected to be nearly four times more expensive than those for TDMA or GSM.[26]

There were other strategic questions to consider, and they all pointed toward a decision in favor of CDMA. The FCC had mandated that wireless operators continue to provide analog access, but there were no GSM dual mode phones. The leading manufacturers – Ericsson and Motorola, who were deeply invested in GSM and TDMA – were not interested in making them.[27] As a result, AirTouch's customers in Los Angeles would not be able to roam on PrimeCo's network. The same would be true for Nynex and Bell Atlantic CDMA customers as well. A decision now to deploy GSM would undermine one of the key strategic reasons for the alliance. Arun Sarin, who was president of AirTouch's international company and a huge fan of GSM, nevertheless believed Schmitt was waging a losing battle. "Between the partners, we had 80 percent of the footprint in CDMA. Why would we want PrimeCo to be GSM?"[28]

The debate over transmission technology came to a head in the summer of 1995, and Lee Cox faced a difficult decision. In June he was joined by the other members of PrimeCo's board, who overruled Schmitt's GSM recommendation and selected CDMA.[29] Frustrated, Schmitt quit weeks later to become the chief executive of another PCS startup, Omnipoint Communications. Schmitt's defection rippled down through the PrimeCo organization as his hand-picked technical staff, many of whom were also GSM champions, followed him out the door.[30] Ginn and Lee Cox were stunned. Though Schmitt had always been a mercurial personality, he had long been a key player at AirTouch. His was the first major defection from a leadership group that had heretofore been remarkably collegial. In the wake of Schmitt's departure, Bell Atlantic inherited the right to nominate PrimeCo's next CEO.

*

For AirTouch, the entire year of 1995 and the first half of 1996 were thoroughly frustrating. The delays with CDMA embarrassed the company on Wall Street. Months after the spin, the Justice Department had challenged AirTouch's assertion that it was free of the constraints of the antitrust decree. If the DOJ could make that stick, AirTouch would lose much of the flexibility it so badly needed. Investors read these signs with concern and seemed to ignore all the good news. AirTouch, for example, had 5.5 million customers worldwide, and its profit margins in cellular were now the highest in the industry.[31] Nevertheless, the price of AirTouch's stock remained flat.[32] Two years after the IPO, the share price had barely surpassed its post-IPO high, despite a rising stock market.[33] The company was plowing tremendous amounts of capital into its networks at home and abroad after winning new licenses in Italy, Spain, and South Korea. Yet Wall Street was oblivious to increases in the company's revenues and subscriber base. In Germany alone, Sarin estimated that AirTouch's share of the D2 joint venture was worth $5.2 billion. But the market valued the whole of AirTouch at only $15 billion.[34]

Around the table in his policy group meetings, Ginn and his top executives began to worry that cellular share prices had peaked. The conversation astonished Mike Miron, AirTouch's new head of corporate strategy. On his first day of work in April 1996, having just uprooted his family from the East Coast, he could not believe it when Ginn asked a policy group meeting, "Is it time to sell?" Miron sensed that Ginn, still a touch too cautious in the Bell System style, saw more risk ahead than he was willing to take. Miron was encouraged, however, to see that other members of Ginn's inner group, particularly Sarin, were aggressively making the case that AirTouch was just getting started. Employee enthusiasm, despite the troubles the company faced, was high and climbing. For his part, Mike Miron concluded that it was impressive "that a company whose mission is cellular could actually put that question on the table for free debate."[35]

As the conversation progressed, all of the central strategic issues facing AirTouch surfaced. Was the pure-play strategy still the right bet? Could the company thrive in regional markets, or did it need to pursue a national footprint more aggressively? Could AirTouch create a national network by affiliation, as it had begun to do with PrimeCo? And what about the global strategy? Should the company consider spinning off the international business to induce investors to focus on its value? If consolidation was inevitable, who would be the hunter and who the hunted?

From the time of the IPO, Ginn had thought about doing a deal with someone. He had entertained casual overtures from several of his peers in the business, but he had not been eager to sell. Arun Sarin was even less interested. He still chafed at the conservative financial tradition of the Bell System. He pointed to the operating results, the steady and dramatic increase in new customers, and the improving outlook for revenues and income. He thought the company was leading the industry in terms of the quality of its performance. Unlike many entrepreneurs who have a strong, single-minded vision of the future and cannot tolerate dissent, Ginn was flexible and sincerely interested in opposing views. He listened to Sarin, Cox, and the other members of his staff and decided they were right. For the present, AirTouch would stay the course. None of them had all the answers to the questions facing the company in 1996 – especially that spring, when the telecommunications landscape, both at home and abroad, suddenly became even more complex.

NINETEEN

The Wireless Explosion

For decades, the Communications Act of 1934 had framed the terms of debate for the telecommunications and broadcasting industries in the United States. The law had been offered to Congress during the depths of the Great Depression as an expedient way to consolidate existing federal regulatory authority and to manage the convergence of telephony, telegraphy, and radio into what some called the "communications industry." The act created the Federal Communications Commission and gave the new agency broad authority to shape the structure of this multifaceted industry through control of the radio spectrum.[1] During the next sixty years, the FCC's regulatory system had been battered by the shifting winds of antitrust law and rocked by successive waves of technological change. But the 1934 law had continued to provide the ground rules for managed competition in the largest telecommunications market in the world.

Although to many observers the Communications Act had begun to seem hopelessly out of date, legislative battles between powerful interest groups in telecommunications and broadcasting had stalemated efforts to rewrite the law.[2] In Congress, it was relatively easy during these years to establish a blocking position but very difficult to create the broad coalition needed to pass new laws. This was especially true when the legislation would have an impact on every congressional district in the country.[3]

After the breakup of the Bell System, the Baby Bells made new legislation a top priority. They hoped thus to escape forever the onerous restraints imposed on their businesses by the settlement. But again, interest-group politics made legislative innovation impossible. They were blocked by the long-distance carriers and independent local exchange operators, who feared the Bells would misuse their control of the local exchange. This

183

time, however, the deregulation movement had achieved substantial momentum, making resistance to change increasingly difficult. Finally, in 1996, the logjam broke and Congress approved the first major rewrite of the nation's telecommunications laws in more than sixty years. President Clinton signed it into law in February.

The Telecommunications Reform Act of 1996 represented a major victory for Sam Ginn and the entire wireless industry. For several years, as Congress negotiated the details of the bill, the shadow of increased regulation had hung over the heads of the wireless companies. In presentations to members of Congress and other federal officials, Ginn had argued that instead of subjecting the wireless companies to the complex web of rules that governed the Baby Bells, "the rest of the industry should be more like us – free to compete."

In the market-oriented atmosphere of the 1990s, Congress listened to the many voices calling for more competition. By repealing the 1982 Consent Decree, the new law freed the Baby Bells and other local phone companies to enter the long-distance, manufacturing, and entertainment–video distribution businesses, provided that the Bells opened their local markets to new entrants and sold access at wholesale rates. The legislation seemed to presume convergence in the telecommunications market, assuming that the major competitors would eventually provide a range of bundled telecommunications services – including local, long-distance, and wireless service.[4]

Within three months, however, the new telecom law had produced effects that were unanticipated by Congress and seemed threatening to some observers, including those in the Department of Justice. Instead of promoting competition, the bill spurred consolidation. Ginn found out about the first major development when his home phone rang shortly after 5:30 in the morning on April Fool's Day, 1996. Phil Quigley said, "I knew you would be in the office." He then told Ginn that the Pacific Telesis Board of Directors had just finished a pre-dawn meeting in which they had agreed to sell the company to SBC Communications, another Baby Bell, for $16.7 billion. The deal, according to Quigley, had been in the works for months. But both sides had waited to make sure that Clinton would sign the new telecommunications legislation.[5]

Ginn was surprised, but not shocked, by Quigley's news. In the two years since the spin, Pacific Telesis had struggled under the new regulatory framework. Although the California economy had finally begun to climb out of recession, competition for local toll revenues and the stringent productivity standards promulgated by the new CPUC framework put pressure on earnings. Squeezed on the revenue side while simultaneously trying to make an enormous $16-billion investment to rewire California for the Information Age, Pacific Telesis had been forced to revise projected earnings downward in May 1995. The company's stock dropped sharply and remained depressed through the rest of the year. In the context of the late 1990s, SBC's interest in Pacific Telesis also made sense. The SBC chairman (Ed Whitacre, Jr.) intended to grow his business into a nation-wide competitor, and he had been outspoken in his efforts to free the Baby Bells from federal restrictions.[6] The acquisition gave him an opportunity to achieve economies of scale and a broader base on which to build his growth strategy.

As one might imagine, the merger aroused intense opposition in Cali-fornia. When news of the deal hit the streets, California politicians reacted angrily, sure that San Francisco would lose a headquarters operation. The media expressed fear that SBC's Texan executives would fail to invest ad-equately in the Golden State's crucial telecommunications infrastructure. In the ensuing media frenzy, Ginn suddenly found himself in the spotlight. A front-page article in the *Wall Street Journal* accused him of having taken "the sizzle" from Pacific Telesis's stock. "They freed up AirTouch to fly," said a senior executive at a rival Bell company, "but they left PacTel to sink."[7] Consumer advocates cried, "We told you so!" to the CPUC, claim-ing that they had predicted Pacific Bell's financial crisis at the time of the spin-off.

Ginn was incensed and frustrated by the coverage. He was, of course, not happy to be portrayed as an avaricious, scheming executive. He felt the press ignored the solid business reasons for the spin. Unwelcome as they were, the accusations proved the wisdom of the decision to take the wireless business out on its own. Had it remained with Pacific Telesis, reg-ulators almost certainly would have been reaching for wireless cash flows to cross-subsidize Pacific Bell's modernization. This was exactly the sce-nario that Ginn had feared. Shareholders would have been the losers. But Ginn's critics weren't buying this argument.

Responding to the *Wall Street Journal* story, Ginn and his communi-cations director, Amy Damianakes, flew to New York to meet with the *Journal* editors. Damianakes had warned him that a protest was unlikely to do any good. But her boss, who thought his personal integrity was at stake, was determined to do *something*. He encouraged the paper's execu-tives to rethink the coverage they were giving to wireless in general, and he gave them a point-by-point rebuttal to their story.[8] The spin, he said, had clearly benefited shareholders. As he pointed out, Pacific Telesis shares were trading at $38 when the spin was announced in April 1992. Since that time, the combined value of one share of AirTouch and one share of Pacific Telesis had risen 82 percent. This appreciation compared favorably with a 99 percent increase in a composite index of the other Baby Bells and 79 percent for the S&P 500 in the same period.[9] He continued: because of the spin, AirTouch had been able to raise sufficient capital to build cellular networks at home and abroad, and Pacific Bell had been able to acquire valuable PCS licenses in the auction. The California licenses – combined with the potential of the California market – made Pacific Telesis attractive to SBC, whose chairman and CEO recognized that economies of scale would be enormously important in the coming telecommunications and cable wars. Of course, Damianakes was right; Sam's lecture didn't change the *Journal*'s editorial stance. But it apparently helped cool Sam Ginn down in the aftermath of the media free-for-all prompted by the SBC–Pacific Telesis merger.

Within weeks, two more Baby Bells announced that they were combining their companies. On the East Coast, Nynex had been facing problems and had suffered a $394.1-million loss in 1993 as it struggled to restructure.[10] Freed by the new telecom law to chart its own course, Nynex set out to merge with Bell Atlantic in a deal valued at $23 billion. Like SBC and Pacific Telesis, the two East Coast companies hoped that greater scale would allow them to reduce costs and leave them better prepared for competition.[11]

With the emergence of these telecom giants, AirTouch – whose market capitalization was just under $15 billion – began to look more and more like bait fish in a pool of sharks. On the road to meet with investors and analysts in 1996, Ginn and his fellow executives stressed the strength of their business, the growth in subscribers, and their increased revenue. But everywhere they went, they looked into stony faces. As AirTouch's CFO Mohan Gyani complained: "We were dogged with negativism, despite the fact that we never missed meeting investors' expectations on operating results."[12] Wall Street continued to lowball AirTouch's international businesses because most of these assets represented minority positions in joint ventures. In some ways, Wall Street had yet to catch up to the new strategy of globalization that was characteristic of many firms, including AirTouch, in the Third Industrial Revolution. The old multinational model relied on direct foreign investment in wholly-owned subsidiaries run by corporate managers. The new model, exemplified around the world in wireless, relied on strategic partnerships that leveraged local political and marketing expertise and financing with foreign technology, operations expertise, and additional capital from overseas. This new model was well suited to a world in which economic nationalism was gradually giving way to a global perspective. Although wireless companies quantified their stakes in these foreign partnerships in numerous ways (e.g., reporting proportional shares of revenue, subscribers, and POPs), Wall Street remained skeptical, at least through 1996.

One good reason for economic conservatism was the failure of many high-profile cross-border strategic partnerships to produce favorable results. For example, MCI and British Telecommunciations created Concert, a marketing and equity alliance, to build a global brand for long distance. But the alliance, which BT had hoped would lead to a merger, foundered on regulatory issues, losses at MCI, and disagreements between management and shareholders. Global One, a similar international brand created by Sprint, Deutsche Telekom, and France Telecom, also struggled – apparently because none of the companies was willing to compromise its own business strategy for the sake of the joint venture.[13] Understandably, these problems tugged at the minds and billfolds of investors.

To counter this investor skepticism and to carry forward their strategy of growth, AirTouch's executives would have liked to increase their stake in most of the markets where they operated. But there was often little they could do. Many governments, including the United States, still limited the amount of investment that foreign companies could make in domestic telephone operations. Nationalism was yielding ground, but it was still a powerful ideology. National security concerns kept popping up, as did social policy. The telephone system was widely regarded as an essential social service and telecommunications policy as a means to achieve social objectives. These ends might be compromised by foreign ownership. Deregulation, privatization, and competition in wireless fundamentally challenged these beliefs and policies.

Ginn had gone to Congress in 1995 to try to change U.S. laws related to foreign investment in domestic telecommunications companies. As he told members of a House subcommittee, these limits were hurting American companies. "What I hear time after time is, 'You can only own up to 25 percent,'" he told a House subcommittee, "'because that's the limitation that your government places on our companies.'"[14] These limits hurt Americans particularly because the Baby Bells were world leaders in the effort to export wireless know-how. While they might provide a joint venture with 100 percent of the intellectual and managerial resources as well as a healthy share of the money, they were limited to a 25-percent share of the equity. These limitations essentially devalued their capabilities, something that was especially troublesome to Ginn because of AirTouch's share price.[15]

But it was hard to buck Wall Street, where analysts were predicting a wireless price war that no one could win. Meanwhile, the pessimists said, endless cycles of reinvestment in the network threatened to diminish even further the returns to investors. Chief strategist Mike Miron worried that AirTouch was creating "penniless prosperity" – by building systems and generating cash flows but constantly recycling these cash flows into new investment. Ginn constantly tested the thinking of his top advisors on the questions of scope and scale. Sarin pushed back. As the industry matured and networks were completed, he said, marketing costs would become a large portion of the company's overhead. A larger footprint would create economies of scale in marketing and help build brand identity. Given that the cost of subsidizing telephones to make them affordable to customers constituted a major portion of the cost of customer acquisition, volume purchasing would yield further efficiencies. Still, the internal debate over strategy continued.[16] As the market shifted from high-flying business customers to individual consumers, average revenues were falling. And the gold-plated customers were often the most likely to jump ship if they experienced quality problems with the network or found more attractive prices elsewhere.[17]

With AirTouch struggling to stay ahead of a rapidly developing market, the shadow of CDMA hung over the company's future. As other cellular providers rolled out digital systems, AirTouch kept promising a new service it couldn't deliver.[18] In these tense times, Ginn received solid support and good advice from his Board of Directors. At Pacific Telesis, the Board had represented the main, old-line companies of the California economy.[19] In 1993 Ginn had purposely recruited a board dominated by entrepreneurs: Donald Fisher had created The Gap from scratch and built it into one of the most successful retailers in the country; Charles Schwab, a failed rock-concert promoter and sometime stockbroker, had jumped into discount banking as deregulation began to take hold in the securities industry and essentially defined this huge new market; Carol Bartz, who had come of age professionally in Silicon Valley, had rejuvenated Autodesk, the leading developer of computer-aided design software; Arthur Rock, the venture capitalist, had played godfather to Intel and Apple. All had been through periods when their fledgling businesses were growing like crazy but failing to be appreciated by Wall Street. From the top, they leaned against the bureaucratic culture of the Bell System and toward entrepreneurship geared to the long run. "Eventually," Fisher told Ginn, "the market will catch up. In the meantime, stay focused on growing the business."[20] But that strategy assumed that AirTouch would be able to hold its own against the competition while its value caught up with its accomplishments.

The competition hit hard in the first half of 1997. In the United States, Pacific Bell and Sprint PCS both launched service in San Diego at the tail end of 1996, starting a price war that signaled the shape of things to come all across the country.[21] At the time, AirTouch's average price in San Diego was four times higher than that of its competition. This had provided a comfortable but vulnerable profit margin. In response, AirTouch at last started commercial service on its own digital network and quickly lowered prices. PCS PrimeCo launched its service in sixteen major cities, covering one fifth of the nation's population, in November 1996. PrimeCo's pricing strategy, which looked to a mass market, eliminated peak, off-peak, and weekend/holiday rates and distinguished only between in-state and out-of-state calls. As this strategy indicated, the company was ready for competition and growth in the consumer market.[22] The battle was on. As Sprint PCS's CEO Andrew Sukawaty told an industry conference that spring, "Having a wireless license used to be a license to print money, and it's not that way anymore."[23]

AirTouch also faced a new competitive situation abroad. Changes on the international front made it evident that AirTouch's core competitive advantage in the race to win new licenses – its experience and credentials abroad – would no longer carry the company across the finish line. When the government of Brazil opened the country to competition in cellular, the change in the international landscape became obvious. Brazil's state-run cellular telephone company had offered to accept an additional 55,000 new customers in the summer of 1996, but it was overwhelmed when more than 1.4 million people applied for service. The government responded by deciding to auction ten regional cellular licenses. The plan touched off a feeding frenzy among the world's largest wireless companies. As AirTouch's Ron Boring told the *Wall Street Journal,* "There's nothing else that compares to this. It's like 10 different countries, with 15 million in population" each.[24]

Bidding for these Brazilian licenses, however, made everyone nervous – and with good cause. Market data was sketchy. Some indicators pointed to a significant upside potential, but Brazil's economy was in transition after a difficult era when inflation rates had peaked at 50 percent *per month.* Inside AirTouch there were fierce debates over how high the company and its partners, including a newspaper and Odebrecht SA (a mammoth Brazilian construction company), should go. When the bids were announced, Ginn and the rest of his team were blown away by BellSouth's winning offer of $2.5 billion, which was nearly twice AirTouch's bid.[25]

The message was clear. Winning new licenses now was all about your stomach for risk and your calculations about scale. Ginn and Sarin concluded that growth by expansion into emerging markets would be much more difficult for years to come. The company would now have to shift its strategy to focus on serving existing markets, increasing the customer base, growing the average revenue per customer, and containing costs. If they were successful, they might be able to slow the steady decrease in margins that came with increasing competition. But none of this would happen overnight.

Investors reacted to intensifying competition as they always had. When Sprint pushed into San Diego, it sent AirTouch stock into a tailspin.[26] Through the first quarter of 1997, the company's stock continued to slide. As it dropped below the IPO level of $23 a share, AirTouch managers quietly stopped posting the daily share price in the halls. Morale sank. The "two by four" goal of doubling the share price within four years of the IPO seemed out of reach.

*

The crucial year of transition for AirTouch and the entire U.S. cellular industry was 1997. Facing greater competition, PCS operators and cellular incumbents revved up their marketing programs to attract subscribers to their new networks. Concerned, Ginn and the rest of his team now took another long look at their organization and practices. Clearly, wireless had the potential to increase the average minutes of use per customer by capturing some of the wireline market. At the end of 1996, for example, the average cellular customer talked for approximately 102 minutes per month. In contrast, the average wireline customer was on the phone 1,000 minutes a month.[27] But industry studies showed that wireless use would not compete with wireline until voice quality as well as prices were roughly comparable.

AirTouch, they decided, had to focus on differentiating customer groups: business users, youth, and so forth. Women were entering the market in droves. In Canada, for example, they accounted for only 17 percent of Bell Mobility's customers in 1991 but by mid-1994 made up 28 percent of the company's subscriber base. AirTouch had to be prepared for the fact that the growing consumer market shifted the demand cycle. Suddenly the pre-Christmas holiday season accounted for a bigger percentage of sales.[28] They needed to increase their emphasis on marketing and to continue reducing costs.

One of the objectives was to cut the cost of acquiring customers. That meant increasing the efficiency of their marketing and retaining customers

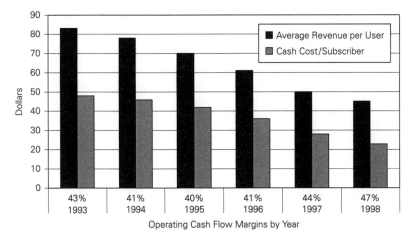

AirTouch U.S. cellular proportionate average revenue per user and cash cost per subscriber. *Source:* AirTouch Communications, "1998 Fact Book," p. 13.

once they had signed them up. In the mid-1990s, U.S. cellular companies were spending an enormous amount of money to win customers – almost $600 per customer.[29] Half of this marketing expense subsidized the cost of the mobile phone. A substantial portion of the rest paid for commissions to retail salespeople.[30] To cut marketing costs, AirTouch had to integrate downstream, strengthening its sales channels and marketing directly through its own stores and telephone and Internet contacts. It also had to reduce the cost of the handset subsidy.

To hold down marketing costs, AirTouch focused on reducing "churn," the pattern of customers switching back and forth between providers. This meant improving service quality and strengthening customer support. Speaking to employees in 1997, Sarin emphasized the importance of retaining high-value customers. "The value of a gold customer," he said, "approximates the value of five new customers."[31] As the number of players in the market expanded, Sarin said, this capability would differentiate AirTouch in the market.[32]

As the firm focused increasingly on cost management and improved operations, Sam Ginn made a critical decision. The business was fast becoming a global organization, but he was still the only officer with responsibility for both domestic and international strategy. True to his Bell System training, that situation left him worried about managerial succession. AirTouch, he thought, needed a clear number-two executive – a president and chief operating officer. As the leaders of the domestic and international companies, Lee Cox and Arun Sarin were both clearly candidates for the job. In December 1996 Ginn and Sarin traveled to Europe to tour the company's operations. Ginn reflected on Sarin's background and training, his ability to move easily between international and domestic venues, his close attention to detail, and his potential as a leader. Shortly after they returned to California, Ginn walked into Sarin's office, closed the door, and said, "I want you to be the COO and run the day-to-day operations." Sarin was delighted, especially now that they had some grounds to be optimistic about the immediate future.[33]

✳

Around the world, increased competition in cellular services was narrowing margins but was also spurring penetration and usage. A 1995 report from the Organization for Economic Cooperation and Development (OECD) showed that subscriber growth rates in countries that allowed open competition were nearly 150 percent greater than rates in duopoly markets. Growth rates in duopolies were, in turn, nearly three times greater than in monopoly markets.[34] As the evidence came in and as more governments turned toward competition, AirTouch began to demonstrate just how efficient its operations had become.

The payoff started to show up in early 1997. New competitors had stimulated demand across the board, and incumbents like AirTouch turned out to be the primary beneficiaries. This was certainly the case in Germany, and it was also true in many U.S. markets. The best news was financial. During the previous year, AirTouch had doubled its earnings per share as cash flows from international operations had finally turned positive and contributed more than $200 million to net income. After dropping to $23 a share in December 1996, AirTouch's stock finally began to soar, nearly doubling over the next twelve months to hit $42 a share by December 1997. The concomitant increase in AirTouch's market capitalization couldn't have come at a better time because, once again, Bell Atlantic was on the prowl.

Having completed its acquisition of Nynex in August 1997, Bell Atlantic had become the largest of the regional Bell companies and the dominant telecommunications provider in the Northeast. It now had customers in fourteen states and the District of Columbia. Under the terms of the Nynex acquisition, Ray Smith (Bell Atlantic's CEO) remained as chairman and CEO of the combined company, to be succeeded by Ivan Seidenberg (Nynex's CEO) one year after the close of the deal.[35] Over the years, Ginn had discussed possible deals with Ray Smith. In the PacTel days, he and Smith and Sprint CEO Bill Esrey and Bill Ferguson, CEO of

Nynex, had met at Pebble Beach and talked about merging their wireless businesses. But the conversation never developed into anything. In Ginn's mind, Smith was more focused on issues surrounding cable and convergence than wireless.[36] But off and on there had been other overtures. In 1994, the threat of a move by Bell Atlantic had been serious enough to include in the PrimeCo partnership a standstill agreement that prevented either one from making a run at the other.[37] But it was not Smith who showed up on Ginn's doorstep in the fall of 1997. It was Ivan Seidenberg and another Bell Atlantic executive named Lawrence Babbio, Jr.

Seidenberg was the future of Bell Atlantic. At one point early in the merger, the press had speculated that he might not survive Smith's regime; Smith had even told *Fortune* that "succession is not ironclad." And Seidenberg had responded to the flap in the press by complaining: "Everybody expects the CEO of Nynex to have to repeat the sixth grade."

Now, at 52, Seidenberg had been CEO of Bell Atlantic for less than six months. A Bronx native with a working-class background like Ginn, Seidenberg had entered the telecommunications industry in 1966 as a cable splicer for New York Telephone. He interrupted his telephone career for a tour of duty in Vietnam, after which he steadily rose through the ranks of the Bell System. With the breakup of AT&T, he went to Nynex, where he became CEO in 1996. With his blue-collar background and his sly sense of humor, Seidenberg enjoyed good relations with Nynex's unions. Wall Street sometimes criticized him for not making deep enough cuts in the company's bloated labor force, but cost-cutting had never been his major claim to fame. That was certainly not what was on his mind when he and Babbio visited Sam.[38]

Larry Babbio wore a number of hats at Bell Atlantic. A lean and intense man, Babbio was a Bell-head with some special talents. He had spent his entire career in the System, starting with New Jersey Bell Telephone in 1966 and rising through the ranks on the East Coast. After the breakup, he went to BellCore briefly and then joined Bell Atlantic in 1987 as vice-president for operations and engineering. Over the next ten years he served as president of Bell Atlantic Mobile, oversaw the company's developing operations abroad, and was promoted to chief operating officer. With the Nynex merger, Babbio became a vice-chairman, responsible for the company's network operations as well as its Global Wireless Group.[39] He was a tough and irascible negotiator. According to one AirTouch negotiator, "Babbio played hardball – with steel balls."

After an initial meeting, negotiations over a possible combination between Bell Atlantic and AirTouch progressed over the next couple of weeks during sessions in Dallas. Both sides recognized that if they could put together Bell Atlantic's East Coast wireless coverage with AirTouch's dominance in the West, they would have the number-one wireless company in the United States. Bell Atlantic was also interested in AirTouch's European assets (they were partners in Omnitel of Italy), but these properties were of lesser importance to Seidenberg. He and Ginn let their staffs work on the mechanics of a deal, but they reserved issues related to ownership, price, and future management to themselves and a few of their senior executives.

At Thanksgiving, Ginn, Arun Sarin, and an attorney boarded the company jet and flew to Fort Worth, Texas, to meet Seidenberg, Babbio, and the Bell Atlantic team. The group made progress but could not agree on how a combined company would be run. Ginn and Seidenberg both realized that AirTouch had the bigger wireless operation, but Bell Atlantic had a larger market capitalization. At one point, negotiations reached a dead end, and Ginn said, "Fine. If we can't settle all of these other issues, take us out." He offered a price for an outright acquisition. Bell Atlantic countered, and the two sides came within a dollar per share of agreement. But then, Babbio wouldn't go any higher. Ginn and the other AirTouch executives weren't sure if Babbio was really this rigid, or if he and Seidenberg had decided to play "bad guy, good guy." Babbio nitpicked over definitions, and it became clear that he would give away nothing that was not spelled out, very precisely, in the contract. He and Sarin battled for their respective corporate interests. Seidenberg seemed to enjoy "the bishop's" role, stepping in when necessary to bring Babbio and his colleagues back into line. But in the end, the two sides went home empty-handed.[40]

Given subsequent events, Bell Atlantic's failure to budge looks like an amazing business blunder. Not as bad as Western Union's nineteenth-century decision not to buy the Bell telephone patents for $100,000. But in the same league. Certainly Ginn and the rest of the AirTouch group were shocked that Bell Atlantic would refuse to go such a short distance to close such a promising deal.[41] The failure made their path to a national footprint a little uncertain. But Ginn and his COO were willing to be patient – and besides, there were important developments around the world that demanded their attention.

TWENTY

Globalism Realized?

Though the digital revolution in wireless – the so-called second generation – seemed to be just beginning in 1997, wireless engineers, regulators, and trade officials already had their eyes on the future. The transition from analog to digital had helped usher in the consumer era in wireless and started a competitive struggle between cell phones and wireline for voice traffic. The transition to the third generation, which promised an era of hand-held video-capable devices with high-speed wireless connections to the Internet, was likely to be even more earthshaking to the industry and society. The Negroponte Flip had predicted that interactive voice would migrate to wireless and data while broadcast television would go to wireline, with an accompanying shift in spectrum allocation.[1] But improvements in wireless transmission technologies were coming so fast that competition threatened to be head-to-head across the market. The central issues that remained unsettled, once again, were whether there would be a technology standard and, if so, which standard it would be.

Sam Ginn and the other executives at AirTouch were deeply concerned about who would make this decision and what, exactly, would be decided. There had been little effort to coordinate a global standard during the second-generation transition to digital. In the move to a third generation – or "3G," as it was known in the industry – the AirTouch crew was determined to do all they could to keep the same thing from happening again. They considered this to be incredibly important. When the industry's technologists were beginning to work on 3G, AirTouch had the greatest global reach of any wireless operator and thus the greatest self-interest in a global standard.

It would not be easy to influence the process of standard setting. In the past, equipment manufacturers and regulators had worked together to define national and international standards. Network operators had taken a back seat. In the precompetitive era, when equipment manufacturers operated in captive markets and there were few interconnection issues (except for long distance), the setting of international standards had been a rather gentlemanly process. But as the equipment business became competitive, intellectual property and the control of technology standards became a key source of competitive advantage for manufacturers and a more important issue for industrial policymakers.

As a result, standard setting ran counter to the general trend toward globalism. This important aspect of technological progress began to migrate from the international to the regional or national level, where political power was concentrated. In Europe, the European Telecommunications Standards Institute (ETSI) became the major player. In the United States, it was the American National Standards Institute (ANSI). At the global level, standards "coordination" replaced standards "setting."[2]

Ginn and AirTouch were left trying to reverse this formidable trend. They all realized that if they weren't heavily involved with the process then the equipment manufacturers would get into a fight and, in all likelihood, government trade officials would weigh in to protect their national champions or regional interests. Once again, consumers and operators would find themselves without a common standard.[3] It would be difficult to prevent a bitter struggle, because telecommunications markets had begun to play an important role in global trade policy and politics. Historically, the international telecommunications organizations had remained relatively passive about the structuring of markets. Instead, they had concentrated on sharing information, developing common standards for shared services, and encouraging technological progress generally. Deregulation and privatization, however, were accompanied by the creation of transnational coalitions of multinational corporations, like Vodafone and AirTouch, that sought to secure and protect economic opportunities in the standard-setting process. These organizations mobilized government support in their efforts to liberalize trade in telecommunications services under the World Trade Organization (WTO) agreement that came into effect in February 1998.[4]

Just as Ginn and the AirTouch team had predicted, a battle over 3G began and quickly evolved into a slugfest between the various manufacturers. As CDMA's superiority became apparent, Japanese and European manufacturers began to look for ways to design around Qualcomm's patents. It was almost impossible. At the same time, policymakers in Europe worked feverishly to find a way to lock out the U.S. equipment makers, who had a six- or seven-year lead on CDMA development. European Union industrial policymakers wanted to protect the substantial market positions developed by Ericsson, Nokia, and other European equipment makers. The war soon spilled over into the International Telecommunications Union.

Craig Farrill and Sam Ginn were desperate to be peacemakers, believing AirTouch had everything to lose if the battle resulted in a divided world. Working through the Clinton Administration's trade representative, they brought pressure on the European Union. But it was hard to get the Administration to mount a vigorous campaign. The government's initial impulse was to let the market decide on a standard. Meanwhile, Japanese manufacturers were sitting on the fence. They had some feeling that they had missed the boat on second-generation handsets and equipment, and they very much wanted to be players in the next round. But initially they, too, were uncertain who would win this struggle.

The battle became even more intense after Qualcomm filed a lawsuit against Ericsson in a Texas court, claiming patent infringement. As the lawyers went to work, Farrill and Ginn intensified their own efforts. They enlisted allies, particularly Canada's Bell Mobility. "What we realized," said Ginn, "was that the only way to solve the problem was to bring the carriers together and announce to the manufacturing community, 'If you don't come together on a worldwide standard, then we're not going to do business with you.'"[5] As the coalition building continued, AirTouch brought into its alliance China Telecom, Vodafone, NTT, Sprint, and all of the AirTouch partners in Europe.

Ginn kept the pressure on both Ericsson and Qualcomm. He suggested to Ericsson that it resolve the dispute by buying Qualcomm. He urged Irwin Jacobs to cut the licensing fees Qualcomm demanded. AirTouch arranged a meeting between Jacobs and Sven-Christer Nilsson, Ericsson's CEO, but the two sides got nowhere.

While Ginn and AirTouch were struggling to be global peacemakers, they had to keep turning away from diplomacy and back to their business at home. Increased competition was quickening the pace of innovation. The competition was repackaging and rebundling services in an effort to appeal to different markets and market niches. Although AirTouch had reassessed its strategy only a short time ago, it found it necessary once again to review its strengths and weaknesses. The AirTouch team believed there were still plenty of opportunities to leverage its well-established capabilities in engineering, global partnering, and operations. But there was a growing perception that some of the organization's strengths were actually beginning to turn into weaknesses.[6]

Coming out of the spin-off, all of the firm's management had embraced the goal of creating a consensus culture that would empower employees and make them feel a sense of ownership. This had been extremely important to Sam Ginn. But now they all realized that the pace of change was accelerating. As the team evaluated the organization in 1997, there was concern that decisionmaking was being slowed by the consensus approach. They decided that they had to centralize some authority without losing the advantages of decentralization – no small task! Centralization would enable them to speed decisionmaking, but they set out to achieve that objective without undermining their employees' sense of ownership or reducing the ability of front-line employees to respond to customer needs.

This became a major campaign within the organization. Hundreds of employees were mobilized in an exercise that aimed to rewrite the company's fundamental value statement. From that process several overriding concerns emerged. The employees who were involved clearly recognized that decentralization was handicapping the company by fostering a lack of cooperation between markets, by creating inconsistent products and processes, and by blurring the focus of the organization. Employees involved in this massive exercise called for more global teamwork and sharing of best practices. They saw the need for a stronger national brand and for more operational synergies between markets.[7] These suggestions were incorporated into a new value statement that became known as the "Compass." Early in 1998, Ginn and Sarin hit the road to begin preaching this new gospel. Their timing was just right.

By the spring of 1998, the stock market – and particularly technology stocks – was roaring upward. AirTouch's position looked better and better. Revenues from the company's international operations were surging. Its position abroad looked very strong. Although some analysts continued to point to the company's minority position in each of its European joint ventures as a liability, others noted that in each partnership AirTouch had reserved the right to select the chief technology officer and to exercise veto power over the marketing director. In that way, it had control over the design of the network and a powerful voice in the way the service was sold.[8] Strength abroad drove the company's stock price, which in early 1998 reached nearly $60 a share – a 260-percent increase over the 1993 IPO price.[9] In this environment, it was even less clear than before whether AirTouch would be the hunter or the hunted. Bell Atlantic might realize what a blunder it had made. And with the market dumping money into the pockets of wireless entrepreneurs, there suddenly were many possibilities for new deals. To Sam Ginn's delight, events were proving that Arun Sarin's optimism about wireless and AirTouch was fully justified.

TWENTY-ONE

A Surprising Long-Distance Call

In one of his worst British press photographs, Chris Gent looked like a caricature of himself. His glasses seemed too big for his face. His teeth exaggerated his grin. There was an element of surprise in his vaguely avuncular expression, as if he were not used to the limelight. Seeing his image for the first time in March 1997, someone might be perplexed to discover that Gent had just become the CEO of Britain's fastest-growing wireless company.

In person, Gent was gracious, self-deprecating, and even somewhat shy. Behind those brown-framed glasses, his eyes were warm and gentle, but he had a quick and sometimes acerbic wit. His trademark suspenders and striped shirts with white collars suggested that he was a man who enjoyed the image of things yet didn't take them too seriously. One observer said his most striking quality was his "dispassionate pragmatism." A member of his board said he was "easy to like and difficult to read."[1] Meeting Gent, one had the sense that the wireless industry was an enormous and fascinating game – one that he played with great intensity. He loved the strategy, maybe even more than he enjoyed cricket, and he was indeed a very ambitious man. As his former boss and predecessor, Sir Gerald Whent, later told the *New York Times,* "I didn't believe he had the verve tycoons are made of, but he has proved me wrong."[2]

Gent had many things in common with other wireless entrepreneurs. Born the son of a sailor in Gosport (a coastal town in the south of England), Gent grew up, like Sam Ginn, in a working-class household. One of four brothers, he and his family moved to South London in the 1950s. His school overlooked the famous Oval cricket ground, and the view fueled his passion for the sport. After his father died while he was still a teen-ager, Gent abandoned formal education and became a trainee bank manager at

Chris Gent – In San Francisco shortly after the announcement of Vodafone's acquisition of AirTouch, Gent told employees that the combined entity would be the first global wireless enterprise.

the age of 19. Hoping to enter politics, he joined the Young Conservatives and was elected national chairman at the age of 29. During these years, he befriended another young conservative, John Major, who would eventually succeed Margaret Thatcher as Prime Minister. Gent toyed for a time with the idea of running for Parliament, but he decided to focus on his budding career in business instead.[3] By that time, he had already demonstrated that he was smart, driven, and socially skilled – a man who might only need one big break to make it to the top.

He got that break in 1984. Then a computer industry executive, Gent was approached by a headhunter on behalf of a startup company in mobile communications. Skeptical at first, he read studies of the industry's potential compiled by British Telecommunications. He liked what he read, recognizing the opportunity to create a business that could make big money. He joined Vodafone on January 1, 1985, the day the company transmitted its first cellular phone call from Trafalgar Square in London to corporate headquarters in the small market town of Newbury, about sixty miles away.[4]

Vodafone was headed by Gerald Whent, who ran the radio subsidiary of Racal Electronics, a defense contractor. In a joint venture with the Swedish firm Millicom, Racal had parlayed its proven engineering capabilities into a winning bid for a wireless license in December 1982. The bid included a bullish marketing forecast, nearly double what British Telecom had predicted for the industry.[5]

As CEO of the new venture, Whent had chosen the location for the company's operations. Newbury, in the Berkshires, had once been a busy inland port on the old Kennet and Avon Canal. The town's High Street crossed over a bridge just below the lock on the canal. The pinnacles to the tower and battlements along the roof of St. Nicolas Church commanded a view of the Georgian and Victorian facades of the town's two- and three-story buildings. Despite its historic charms, Newbury was struggling in the mid-1980s and could offer good deals on office rent. Whent settled his new operation inside a courtyard near where High Street came to a T with the London Road.

During 1983, Racal was engineering and then constructing its cellular network, using Ericsson to supply the equipment. As construction progressed, Whent took a wireless tour of the United States. Of all the companies he visited, PacTel was the most impressive. Phil Quigley was candid about L.A.'s demand and gave Whent confidence that the cellular business might actually be profitable.[6]

When his network was being tested in late 1984, Whent hired Saatchi and Saatchi to come up with a sexy name for the new company, something that would be more appealing to consumers than Racal Telecom. Hoping for the best, Saatchi and Saatchi wanted a term that could become generic – like Kleenex or Xerox. Vodafone seemed to offer that possibility. It combined with phone the idea of "voice" and "data" in the first two syllables, offering a universal term for wireless service.[7]

With Vodafone's launch of commercial service in 1985, industry analysts predicted the company might win 40 percent of the domestic cellular market, with British Telecom's Cellnet remaining in the dominant position. In the first year, that prediction came true. But the size of the national market surpassed both companies' expectations. Within a year, more than 100,000 customers had subscribed for cellular service in the United Kingdom, three times what Racal had predicted in its marketing study.

Fortunately for Vodafone, British Telecom had made the same critical error that the Bell System had made in planning its original cellular operations: it had grossly underestimated demand and built a thin system with insufficient capacity. Customers had trouble placing calls, a problem that flashed "opportunity" to Whent. Vodafone built its system with twice the capacity of Cellnet and, as a result, the company's customers experienced fewer service problems. Even then, early growth threatened to overwhelm the network, leaving Whent facing the same situation in London that Ginn and Phil Quigley had faced in Los Angeles. Vodafone responded as they had, racing to add capacity by building more towers and dividing cells.[8] By the end of the second year, Vodafone had 56 percent of the domestic market, and Cellnet appeared not to have the organizational vigor required to recover its market lead.[9]

Despite Vodafone's success, Racal's stock suffered in the late 1980s as its core businesses declined. The unwillingness of investors to place a high value on the company's wireless subsidiary frustrated Racal's chairman, Sir Ernest Harrison, just as it had Sam Ginn.[10] Frustration turned to alarm in 1988 when Goldman, Sachs called to warn Harrison that Racal was vulnerable to a hostile, lowball takeover bid by Cable & Wireless, which was looking for a cheap way to get into the cellular business. Goldman, Sachs suggested an IPO of some of Vodafone's stock, telling Harrison that Vodafone alone was worth an estimated $2 billion. When Racal announced the initial public offering, the company's shares rose 33 percent in a single day.[11]

Even though he successfully sold 20 percent of the company that fall, Harrison continued to be concerned about the value of Racal's shares. The launch of PCN in 1989 posed an enormous competitive threat to incumbent cellular operators, and investors once again beat down the price of Racal's shares. Harrison finally concluded in 1991 that a spin-off was the only way to realize Vodafone's full value for shareholders and get Racal out from under the shadow of its offspring.

The spin-off, which was completed in September 1991, offered investors a "pure play" in cellular with a fast-growing company. Over the next seventeen months, Vodafone's share price rose dramatically, three times faster than the standard index of shares on the London stock exchange. Within five years, the company's market valuation had increased from £3.5 billion to £7.1 billion, making it one of the hottest stocks in Europe.[12]

Viewed from California, Vodafone's success helped persuade Ginn that the Pacific Telesis wireless operation would create shareholder value and promote the growth of his own business. Ginn also kept his eye on Vodafone for other reasons. Throughout Europe, PacTel and Vodafone were competing for the premier cellular licenses. These were the "second" licenses in national markets where the first license had been issued to the incumbent wireline monopoly or PTT. As both Ginn and Chris Gent – now a rising star at Vodafone – had quickly recognized, these second licenses were the best. Competing against an unreformed monopoly, they could win market share relatively easily on the basis of better, more customer-friendly service. In this kind of duopoly, the monopolists' high cost structure allowed for terrific margins.[13]

Going abroad, Gent, too, had to master the art of the joint venture partnership and learn how to be effective with minority ownership. Like PacTel, Vodafone offered its network management and operational experience to consortia bidding for licenses. It located local partners with strong financial, industrial, and political credentials. Initially, the company's stake in these partnerships ranged from 20 to 45 percent of the business.[14]

How did Vodafone do in head-to-head competition with PacTel's wireless crew in Germany, Portugal, and Italy? The full application package – including experience, technology, market forecasts, and local partners – was of crucial importance at that time, and Vodafone thought it had a competitive edge because it was a European company. But PacTel repeatedly won. The losses were so disappointing that Whent, fearful of diluting Vodafone's earnings, pulled in his corporate horns and retreated from the competition for licenses in Europe.[15] Gent couldn't tolerate this conservative strategy, and when he was picked to succeed the aging Whent as CEO in 1997, he determined to chart a more aggressive course.

For some time, Gent had wanted to lower Vodafone's prices to bring in new customers. He believed that, with the transition to digital, Vodafone's system had the capacity to reach a mass consumer market. He also thought the firm needed to restructure its distribution system to prepare for intense competition. Whent had rejected these strategies, preferring higher margins to greater volume.[16] When the U.K.'s newest digital competitor, Orange Plc, started breathing down Vodafone's neck, Whent had finally agreed to lower prices.[17] As a result, Vodafone was able to develop a range of new plans. New customers who wanted a mobile phone for emergencies could pay a low monthly service charge but higher per-minute call rates. High-volume callers could choose lower per-minute call charges in exchange for a higher monthly minimum.[18]

One of the first things CEO Gent changed was the distribution system. In the early days of cellular in the United Kingdom and the United States, both governments forced cellular operators to act as wholesalers and to rely on a network of dealers to retail phones and sign up subscribers. In part, Vodafone's success derived from its strong relationship with dealers and its ability to keep them happy. In competition with British Telecom, Vodafone established a reputation for better service to the dealer community. With the launch of PCN, however, the government removed the wholesale-only restriction, allowing network operators to establish their own retail systems.[19] Orange Plc went directly to the consumer and was able to build a strong brand very quickly. By eliminating transaction costs – the heavy commissions to retailers – the company was able to lower costs and prices. In this new market, Vodafone's complex network of dealers became a liability.[20]

Gent consolidated the company's service providers into six businesses. He integrated downstream, setting up his own retail division and organizing the division into three functional units that handled the retail, corporate, and dealer markets. He also pushed hard on marketing, introducing a new logo shaped like a single apostrophe to represent the idea of speech. Gent hoped that ultimately it would stand out like the Nike swoosh and become the image of mobile telecommunications.[21] As he consolidated Vodafone's distribution channels and stepped up its marketing efforts, Gent began aggressive price competition to take market share away from Cellnet and Orange. The number of Vodafone subscribers and the firm's revenues began to rise dramatically.[22]

With the domestic business surging, Gent turned to issues of global strategy. The development of a mass market in Europe and in the United States put increased pressure on wireless operators to achieve economies of scale. Gent decided that his company needed either to seek new strategic alliances or make acquisitions in order to get up to global scale. He began to look with interest at AirTouch. The two companies went head-to-head only in Germany.[23] They owned one license in common (Europolitan, in Sweden), and they were partners in Egypt. Everywhere else their properties complemented one another. In addition, both companies had invested in the satellite company Globalstar. In 1997, Gent had little interest in the American market, with its complex sets of metropolitan licenses and growing array of players. British companies, he thought, generally got into trouble when they tried to move into the United States. But it made perfect sense that Vodafone and AirTouch should find a way to combine their European assets and create a regional powerhouse.[24]

Before he could try to put together that deal, however, Gent had to tackle a number of other issues. His chairman, Ernie Harrison (the former CEO of Racal), was not interested in Gent's "grand plan" for world or even European domination. He would have to be persuaded that a major initiative was needed. That wouldn't be easy because, as Gent knew, other and larger firms were interested in AirTouch. Gent was not sure how to outmaneuver them; nor was he sure what it would take to get AirTouch interested in a deal.

Vodafone and AirTouch had danced around each other for years. At industry meetings, Gent frequently found himself on the same panel with Arun Sarin. Watching him speak, Gent developed respect – and even affinity – for Ginn's right-hand man. They swapped notes on the industry.[25]

At a conference in 1996, Gent broached the subject of putting AirTouch's European assets together with Vodafone, either through a joint venture or merger. But Sarin made it clear that neither he nor Ginn had any desire to split the company. They were concerned, he said, that dividing the business would have enormous tax consequences. The two talked about how they might create operational synergies through purchasing or contractual roaming agreements. Gerald Whent, before he stepped down as CEO, had also raised the idea of collaboration with AirTouch. He and Ginn had had breakfast and talked vaguely about the idea of a strategic alliance. But nothing came of this conversation, either. After Gent became CEO, he decided it was time to turn up the heat and aggressively seek an alliance with AirTouch.[26]

His first move was to find out what the other players were doing. For years Vodafone had relied on Goldman, Sachs for investment banking and strategic advice, and Gent now turned to them for help. With Scott Mead from Goldman, Sachs, he traveled to the United States in February 1998 to meet with executives from Bell Atlantic and MCI.[27] Both companies clearly had interests in AirTouch. Gent wanted to find out whether they intended to make a play for the company. If they did, he hoped they would choose to sell AirTouch's European assets to Vodafone.

But the news he received in the United States was not encouraging. In New York, Bell Atlantic's CEO Ray Smith turned out to be more ambitious than Gent had imagined. Having just completed the acquisition of Nynex, Smith said flatly that Bell Atlantic intended to buy AirTouch and then come after Vodafone. He smiled when he said it, but Gent was not charmed. Bernie Ebbers, the CEO of MCI WorldCom, was far more personable, but he didn't hold out any immediate opportunities for Vodafone. Having just acquired MCI, he was not ready for another merger. He told Gent, "I don't want to do wireless, not yet." Down the road, he said, in maybe two years, he envisioned the possibility of a run at AirTouch.[28]

Having scouted out two of the major competitors, Gent organized a top-level strategy meeting in the United Kingdom. During this session, he posed several critical questions. First, was wireless enough? In essence, four years after the fact, Gent was facing the same issues that Ginn had confronted at the time of the spin and continued to face as AirTouch's stock languished on Wall Street. Would the pure-play strategy work? Just as they had in 1993, investors and telecommunications analysts were predicting that the convergence of voice, data, and video demanded that companies be in both fixed and wireless networks. This was even more important, they said, now that the Internet had become a bigger factor in the market. Of course, Gent and his top executives disagreed. They were convinced that wireless as a pure play had a terrific future. Convergence, they thought, was still far down the road. In the meantime, consolidation in the wireless business offered the most appealing strategy.

Europe was the obvious next step for Vodafone, but Gent asked his executives to also think about a global strategy that included the United States. A year earlier he had rejected a move into America, but new data showed that U.S. market penetration rates continued to lag many countries in Europe. According to Gent, that meant "there was lots of growth to come." There would be another generation of wireless down the road, and Gent anticipated the day when there would be a single global standard. "Therefore, we had the potential for a single service worldwide for our users." If this was their goal, AirTouch was even more attractive, and Gent quickly shed his qualms about Brits investing in the United States. With Ernie Harrison's imminent retirement from the board, Gent's internal roadblock would soon be gone as well.[29] By the end of the annual strategic review, a consensus had emerged. Vodafone had to choose either "to hunt or be hunted."[30] As the hunter, Vodafone's number-one quarry had to be AirTouch.

Gent placed a call to Sam Ginn. He eased into the subject by mentioning his conversation with Sarin the previous year regarding the European assets. But now, Gent said, Vodafone was thinking of a merger. Ginn told him that he was not in any hurry to put together a deal, and in any case he was focused on the emerging structure of the wireless business in the United States.

Ginn was just being cagey, and Gent completely understood this type of game. AirTouch had been considering the possibility of combining with Vodafone. Strategically, as Arun Sarin put it, "It was a no-brainer," because the two companies had virtually no overlapping properties. "But every time we did the math, and we looked at how much of the company we would have to give up, we hesitated."[31] AirTouch calculated that, to acquire Vodafone, it might have to let go of as much as 47 percent of its equity. To former Bell-heads, this seemed dangerously close to giving up control. Still, Sam Ginn didn't reject Gent's idea outright. Before he would consider a deal, he said, he wanted to get to know Chris Gent the man. Ginn suggested that he come to California in the summer for the annual encampment at Bohemian Grove, and Gent accepted.

During three days in the California redwoods, Ginn and Chris Gent got to know one another a little better. To Gent – who may have been a little awed by the huge trees, the California culture, and the unique traditions of the Grove – the experience was interesting, to say the least. It was not a business environment, but it did give him the chance to get to know "the quality and substance" of Sam Ginn. Although he found Ginn to be smart and wise, he was also surprised that Ginn was not as gregarious as most Americans. He, too, was a little shy.[32]

Gent was eager to talk about business and a possible combination of their two companies. One afternoon, as they were lying on bunk beds like schoolboys at summer camp, Gent raised the subject. But Ginn held him at bay. The code in the Grove discouraged business dealings, and because AirTouch's combination with U.S. West was still in registration, Ginn was barred from discussing another deal without disclosing this information to the Securities and Exchange Commission (SEC). That didn't prevent them from talking about the general future of wireless. Everyone in the business knew that large-scale consolidations were on the horizon. For AirTouch, the major strategic question was whether to pursue a global approach now or consolidate its position in the United States first. In passing, Ginn told his British guest that AirTouch had come very close to making a deal several months ago. But the conversation quickly shifted to other topics, and small talk kept them occupied as the sunlight filtered through the redwoods and into their cabin.[33]

Not long after Gent left the woods and returned to the United Kingdom, he met with his Board of Directors. The company had a new chairman, the former CEO of Tesco and deputy chairman of Whitbread, Ian MacLaurin. MacLaurin, who had replaced Ernie Harrison in January 1997, was inclined to favor Gent's global ambitions. The group listened to Scott Mead from Goldman, Sachs run through the numbers on a possible merger with AirTouch. Plan B, if a merger wouldn't fly, was to pursue a takeover. Gent didn't prefer this route. It would mean taking a large hit to the company's earnings per share, because Vodafone would have to pay a premium for goodwill.[34]

Merger rumors started flying. On the road that fall to talk with investors and analysts, Gent was bombarded with questions about AirTouch. Analysts wanted to know when the corporate marriage would be announced. To many of them, a merger seemed to be the logical and sensible thing to do. Gent could see that there was considerable momentum growing for a deal. He called Ginn again and said that he would like to come back to California – and this time he would like to bring his team.

*

Their secret meeting took place in a hotel in San Francisco. Gent brought his chief financial officer, Ken Hydon, and the executive who ran the company's international businesses, Julian Horn-Smith. Ginn came with Sarin and AirTouch's CFO, Mohan Gyani. All of these men had met before and clearly respected each other, which eased the tension of what they knew might be a mega-deal. As they sat down on opposite sides of the table, they shared perspectives on the future of the industry. Both teams believed that wireless would emerge as a global business. In the less developed parts of the world, wireless offered a fast and cheap way to build a modern telecommunications network. In the developed world, it had certain intrinsic advantages over fixed-line service, especially for voice communication. Both sides also believed that a sharp focus on wireless was the way to go, but they felt pressured to defend their pure-play strategy against the many champions of bundling, integration, and convergence. This was a comfortable meeting, with an easy rapport.[35]

The next step was not so easy. The Vodafone team had marriage on their minds. "They were in a selling mode," according to Ginn. But AirTouch was not necessarily buying the idea of a merger. Although Arun Sarin had discussed collaboration with Gent before, he now challenged the Vodafone team to make their case: "Tell me why this makes sense."

"Together, we would cover Europe," Julian Horn-Smith responded. "We work well together, and we would be unbeatable." With consolidation on the horizon, he continued, both companies faced the threat of being swallowed by one of the existing telecommunications giants. As liberalization took hold in Europe, the former PTTs – including Deutsche Telekom, France Telecom, and Telecom Italia – were like bears coming hungry out of hibernation. They were searching for companies like Vodafone and AirTouch that would give them global strength. But if Vodafone and AirTouch combined forces, they could become the kind of bear that would be feared and respected in telecom's international hunting ground.[36]

As the discussion progressed, it became clear that the issue of ownership was especially tricky. While AirTouch was a bigger company, with more customers and revenues, Vodafone's stock had a higher multiple. AirTouch had a problem insofar as demand was growing much more slowly in the United States than in Europe. In Europe new services, especially short message service (SMS, a truncated e-mail system), were driving usage. The widely accepted policy of "calling party pays" in Europe also provided further incentive for cellular customers to keep their phones turned on. In addition, the scale economies of GSM made handsets and system equipment cheaper compared to those used in CDMA networks. As a result, margins in the United Kingdom were better, at least for the present.[37]

The AirTouch team was concerned that Vodafone's high-flying stock price might not hold up, particularly in the aftermath of a deal. Accounting rules in the United Kingdom would require Vodafone to amortize goodwill on any transaction. If AirTouch received the premium that Ginn wanted, then Vodafone would have to shoulder pre-tax losses for years. Nevertheless, Vodafone's share price gave AirTouch less leverage on other issues.[38]

Management issues became critical to the discussion. At one point in the conversation, the two CEOs excused their deputies, who took a walk to Starbucks for coffee.[39] Alone together, Ginn and Gent got down to the question of individual roles. Gent proposed that Ginn would become the chairman and Gent the CEO of the combined company. Ginn had no problem with this idea. Although he hadn't planned to retire in the near future, he had already begun to prepare for an eventual transition in which Sarin would step into his shoes. That quickly became a thorny issue.[40]

Ginn believed in the standard American corporate model, which employed a chairman–CEO teamed with a chief operating officer (who was usually the heir apparent). Gent preferred the British model, in which key executives reported directly to the CEO. These differences in managerial philosophy had significant implications for the executives involved and for the possibility of completing a merger. Both Gent and Ginn were loyal to the members of their respective teams and wanted to make sure they were protected. Ginn badly wanted to be assured that Sarin would become COO of a combined entity. Gent was reluctant, with equal feeling, to bring Sarin in over the heads of his loyal team members. He was also concerned that, with Ginn as the nonexecutive chairman and Sarin as the COO, he would be squeezed in the middle. He wanted to continue to run his business.

After Vodafone's team returned to Newbury, there were several phone calls back and forth as the two sides continued to wrestle with these several issues, including valuation. Gent sought some way to resolve the disagreement over management. Just before Halloween, Ginn and Sarin left for Egypt as part of a tour of international operations that would highlight the current state of cooperation and competition with Vodafone. In Egypt, they evaluated the current operations of Misrfone, the joint venture operated by AirTouch, Vodafone, and the company's Egyptian partners. They also took time with Ann and Rummi to visit the Pyramids, the Sphinx, and the mummies in the Egyptian Museum. From Egypt, the team went on to Sweden for an update on Europolitan's operations. Then it was on to Dusseldorf to visit Mannesmann Mobilfunk and have lunch with Mannesmann's CEO, Joachim Funk, and his heir apparent, Klaus Esser. After one more stop in Milan to evaluate the Italian partnership, the two executives flew to London on the morning of November 6.

They met Chris Gent in his offices and took a tour, Ginn's first view of Newbury. Afterward, Ginn and Gent returned to the negotiations, but to Ginn it was clear that they were not making progress. Finally, he said, "You know, we're not getting anywhere. Let's just call it off, and see what happens."[41] Gent agreed. This was not a deal that could be forced ahead. That night, Ginn and Sarin had dinner with Gent at a restaurant called the French Horn along the Thames. It was a casual, comfortable meal. Gent was not disappointed by the failure of the negotiations, and he knew that Ginn had other suitors. He was willing to bide his time for the moment and see what other deal Ginn might put together. If he had to, Gent was prepared to move quickly to trump any agreement that might emerge.[42] Ginn had decided to end the conversation with Vodafone because he had a new proposal coming from another front: Bell Atlantic CEO Seidenberg had returned to the issue of consolidation, except that now the situation for everyone had changed dramatically.

AT&T had dropped a bombshell when it announced a bold new marketing program in May 1998. Digital One Rate allowed customers to pay a flat rate for a prespecified number of minutes, with free roaming throughout most of the United States. Almost overnight, the program sparked fundamental changes in the marketing of wireless and altered the geographic basis of competition. AT&T had made a decisive move to build a national brand and a national service. The plan, which was promoted by AT&T's wireless president Daniel Hesse, was so popular that the company was swamped by demand. Business men and women who traveled frequently and often logged more than a thousand minutes a month on their cellular phones rushed to sign up for the new program. AirTouch, Bell Atlantic, and other big U.S. wireless companies were caught flat-footed.

Most had anticipated the emergence of national wireless brands, but they didn't expect them this soon. AirTouch's domestic acquisition program since 1985 had focused on two key objectives: obtaining valuable cellular assets (in terms of markets) and making strategic expansions of its geographic footprint. Sam Ginn and Arun Sarin anticipated that they would someday have to make a major merger or acquisition if they were going to provide nationwide service. But AT&T had suddenly moved up the time schedule for everyone.

AT&T's innovation also upset the industry's pricing logic. Economists have long argued that customers should prefer measured service plans that allow them to pay only for time that they use on a network system.

Customers, however, are rarely economists. As the history of the Bell System showed, most telephone users did not like worrying about how much they were spending on measured service. They also didn't like multipart rate structures (such as roaming charges). Flat rates were easier for noneconomists to understand. AT&T had learned that lesson the hard way. In the early days of telecommunications, when independent companies had lured customers away from the Bell System by offering flat-rate plans, AT&T had stubbornly refused to follow suit until it had lost significant market share.[43]

Dan Hesse, a 23-year veteran of AT&T, had borrowed a page out of the Bell System's history to score a coup. It had not been easy. Executives in the wireless company had been pushing for a national calling plan that would eliminate roaming charges since shortly after the completion of AT&T's acquisition of McCaw in 1994. AT&T had resisted, fearing that such a plan would cannibalize the company's long-distance business.[44]

AT&T's Digital One Rate plan leapfrogged over the localized market structure created by the FCC in the early 1980s in favor of a plan that fit better with customers' perceptions of what a wireless phone should do. As AT&T recognized, customers could not perceive the invisible boundaries that separated the San Francisco market from San Jose, for example. Nor could they intuitively understand why it should cost them more to turn their phone on and use it in New York than it did when they used it at home in California. Sam Ginn and others in wireless knew that, and they had worked to integrate regional markets like San Francisco and San Jose. Now AT&T had suddenly taken that effort to the national level.

Digital One Rate reflected a major organizational shift underway at AT&T, a development that would have an impact in other wireless operations as well. McCaw's cellular businesses had operated with a decentralized structure in which a great deal of authority was delegated to local managers. This structure reflected the FCC's original decision to license cellular on the basis of metropolitan areas (in sharp contrast to the national licenses in Europe), a decision that encouraged the creation of a decentralized management structure. Decentralization encouraged innovation, but it diminished the ability of companies to derive economies of scale in marketing and to build regional or national brands. One consequence was the high cost of acquiring customers – as much as $300 per head. According to one former wireless marketing executive, increases in subscriptions were driven "by the brute force of sales resources."[45] As AT&T implemented the Digital One Rate program, Hesse also centralized authority in the company, making it a consolidated national organization.

One Rate was an exciting competitive maneuver, but many executives in wireless thought they could see some flaws in the plan. Some were sure that AT&T's promise had overreached its wireless capacity. Indeed, when demand exceeded capacity in New York, the company earned a black eye in the press. More baffling to executives at AirTouch was the fact that AT&T had priced One Rate at such a steep discount. At a dime a minute, AT&T's new rates cut margins in the cellular business from 40 to less than 20 percent. If those prices stuck, everyone in the industry would have to learn how to run more efficiently or else leave the market to AT&T.

Bell Atlantic clearly understood what was at stake. Two months after the AT&T announcement, the firm made a deal to merge with GTE. By acquiring what had once been the largest independent wireline telephone company in the nation, the Baby Bell expanded its national presence dramatically. That merger also promised to make Bell Atlantic–GTE's wireless operations nearly as large as those of AirTouch, and this gave Seidenberg additional bargaining power in his negotiations with Ginn.

When Seidenberg and GTE's CEO Chuck Lee came to visit Ginn in San Francisco in the fall of 1998, they floated a new scenario. Their initial plan was to combine Bell Atlantic, GTE, and AirTouch's wireless operations in a new company. Ginn said that he would consider that sort of deal. The negotiations on the specifics of an agreement, however, dragged on for weeks, and then regulatory accounting issues suddenly emerged as a major complication. Bell Atlantic and GTE were hoping to qualify for pool accounting, which would reduce the tax consequences of their combination. But that arrangement limited their flexibility in structuring a wireless combination. AirTouch also had some accounting issues of its own relating to the acquisition of U.S. West's wireless operations.

It gradually became clear to both sides that only an outright acquisition by Bell Atlantic would work. The two companies decided to seek guidance from the SEC, hoping not to reveal their negotiations to the world.[46] But it was impossible to keep any deal of this size confidential. Although negotiations between Bell Atlantic and AirTouch had not been reported by the press, rumors were already moving through the investment community in the last weeks of 1998. Hearing these reports, Ken Hydon of Vodafone began calling Mohan Gyani at home around Christmas. As Gyani knew, Ginn had made a gentleman's agreement with Seidenberg and Lee that he would not entertain other offers while they were in serious negotiations. Gyani didn't return Hydon's calls. But the silence in San Francisco was telling to Hydon and his boss.

TWENTY-TWO

Bidding for the Future of a Wireless World

In the locker room at Kiele Mokihana, a private golf course on the island of Kauai, Sam Ginn bent over to untie the laces on his golf shoes. As he leaned back to remove his shoes, Ginn looked up at the muted television above the lockers. Startled, he saw the blue and yellow logo of his company, Air-Touch Communications, paired with the symbol for telecommunications giant Bell Atlantic. Without the sound, he couldn't tell what the CNBC newscaster was saying. But he could guess. Somehow news of his conversations with Seidenberg had leaked. Reaching for his cell phone, Ginn stepped outside the locker room. Looking out over the greens to the Pacific Ocean beyond, he punched in a call to San Francisco.[1]

*

At AirTouch headquarters in San Francisco, Arun Sarin's staff had spent much of the morning frantically calling every golf course on Kauai looking for Ginn. When his secretary told him that Ginn was on the line, Sarin lifted the phone and asked: "Where the hell have you been?"

Ginn smiled. "Catch me up," he said.

Quickly, Sarin told him that – after the CNBC report – the phones at AirTouch had lit up with calls. Reporters wanted confirmation of the impending deal and details. "Things are really getting hot," Sarin said. "The jet is on its way to get you."[2]

Sarin was not the only executive searching for a wayward CEO on New Year's Eve. Seven thousand miles away in London, Steve Mead, an investment banker with Goldman, Sachs, had also seen the CNBC report. Mead realized that Vodafone's opportunity to buy AirTouch was slipping away. He and his British client would have to move quickly if they were going to

214

beat Bell Atlantic to the punch. But Chris Gent was on the other side of the world.[3]

An avid cricket fan, Gent had flown to Sydney to watch the underdog English team battle Australia. He also planned to attend the board meeting of Vodafone's Australian subsidiary. When Gent received the bad news about AirTouch and Bell Atlantic, his reaction was characteristically understated. "That's troubling," he remarked.[4]

The next day, Saturday, Gent faxed a one-page letter from Australia to Ginn's San Francisco office expressing his interest in making a deal. Then, with business out of the way, he joined an opening-day full house at the Sydney Cricket Ground. The weather had brightened. On that day, the English team staged a miraculous fight against the Australians.[5] Gent had reason to believe the gods were smiling on the British.[6]

On Sunday, January 3, Vodafone's offer spewed out of the fax machine in San Francisco. The proposal was stunning. It reflected the brazen track record Chris Gent and Vodafone had compiled in Europe. Vodafone topped Bell Atlantic's bid by nearly 25 percent! Where Bell Atlantic had negotiated a stock swap worth $45 billion – valuing AirTouch shares at $72 each – Vodafone offered a combined stock and cash deal worth $56 billion. Gent valued AirTouch at $90 a share.[7] For Ginn, Arun Sarin, and the rest of the AirTouch team, the new bid sent their adrenaline into overdrive.

In fact, Gent's timing was impeccable. A handful of major telecom companies were salivating at the prospect of snatching AirTouch's assets, but in January 1999 they were all encumbered in one way or another. Bell Atlantic continued to have issues with its acquisition of GTE, and these problems forced the company to move even more slowly than usual. This delay freed Sam Ginn to entertain other offers, including Vodafone's high bid. Mannesmann, which might have been interested in doing a deal with AirTouch, was struggling to engineer the separation of its long-time industrial operations from its fast-growing telecom business and hesitated to make such a major move before that project was complete. Over the next week, other hungry competitors would take a long look at AirTouch but, for one reason or another, would shy away.

*

News of the bidding contest between Bell Atlantic and Vodafone broke early the following week. As reporters scrambled to get the inside track on the deal, Vodafone and AirTouch organized teams of negotiators to meet in New York City to hammer out the details of a contract. New York was full of rumors. Other potential buyers, it was said, were entering the auction.

The *New York Times* reported that British Telecom was considering an offer. Ginn took a phone call from Bernie Ebbers – the flamboyant, up-start CEO who had created a telecommunications giant when his company, WorldCom, acquired MCI. Ebbers had no wireless division of his own be-cause MCI had foolishly sold its cellular properties to Craig McCaw for a paltry $122 million in 1986. Ebbers now believed that the lack of a wireless entity would hobble his company in the worldwide race for telecommuni-cations leadership. Talking to Ginn on the phone, he suggested that MCI WorldCom and AirTouch could do great things together.[8]

Ebbers, a former basketball coach, had a remarkable ability to create a conspiratorial climate. Like a good car salesman, he seemed to say, "It's just you and me against the world." Amused, Ginn nevertheless told Ebbers that the price was already on the table. "What are *you* prepared to offer?"[9]

Ebbers promised a quick response. But that week, when the press re-ported that MCI WorldCom was considering a $55 billion offer for Air-Touch, the company's shares plunged 11.5 percent in two days. The largest shareholders feared the dilution that would come if the corporation, still digesting the acquisition of MCI, also gobbled up AirTouch. In a second call later in the week, Ebbers explained to Ginn that he would not be able to join the bidding after all, and MCI quickly issued a press release in an effort to bolster the price of its stock.[10]

Bernie Ebbers was not the only CEO worried about dilution. With Vodafone's offer on the table, Bell Atlantic had to decide whether it could improve its bid. Since the breakup of the Bell system, Bell Atlantic had emerged as one of the most aggressive of Ma Bell's children. With the ac-quisition of Nynex in 1997, the company had staked its claim over most of the local telephone markets in the Northeast. Its territory encompassed fourteen states and the District of Columbia, with telephone lines stretch-ing to 22 million homes. The acquisition of GTE further expanded the company's reach, adding local telephone customers in California, Florida, and Texas. GTE also gave Bell Atlantic a strong foothold in the long-distance business, extensive data networking services, and more wireless customers.[11] But each of these acquisitions diluted earnings, leaving in-vestors leery of yet another purchase. When the stock market opened on Monday, January 4, following the first revelations of a possible Bell At-lantic bid, the price of Bell Atlantic shares dropped. Seidenberg was left to ponder that message. When Wall Street learned of Vodafone's offer the next day, and when the shares of both Vodafone and AirTouch rose, he had more cause for concern.[12]

Pushing ahead despite Wall Street's reaction, Seidenberg and Ginn talked almost every day while AirTouch's negotiators were working with Vodafone on the particulars of a deal. Midweek, when the SEC gave an affirmative ruling on the accounting issues involved in an AirTouch–Bell Atlantic combination, the press speculated that Seidenberg would now raise his offer.[13] He told Ginn that he was talking to his Board. Some of Ginn's own shareholders had turned up the heat on both CEOs by launching a shareholder lawsuit to force AirTouch's Board to take the highest bid.[14]

Chris Gent, now back in England, was getting anxious. He had not talked to Ginn personally about Vodafone's offer and could only guess what Ginn was thinking by the response from AirTouch negotiators. Finally, on Tuesday the 12th, he called. The two men discussed a variety of concerns that had not been resolved by their staffs, including the number of seats each party would have on a combined board, the roles of key executives, and the retention of employees.

Ginn found it much easier to resolve issues with Gent than with Bell Atlantic, and he said so. But despite the progress they had made, Ginn was still not convinced that Vodafone represented the best choice. The British offer depended on its high-flying stock, and there were no guarantees that the share price would stay up. Ginn's fellow executives also sensed that, at his core, he preferred a deal with Bell Atlantic because it was a local exchange company. To a Bell-head or a former Bell-head, that meant permanence. Ginn knew, however, that he had to maximize the value of this deal for his shareholders, and as the bidding progressed, he stayed in close touch with his Board of Directors. He also talked to Seidenberg almost every day. Then, he asked the Board to meet on Friday, January 15, to make a final decision.[15]

Ginn called both Gent and Seidenberg the day before the meeting to give them one last chance to improve their offers. The next morning, at 7:30 A.M., Gent responded by increasing the cash portion of Vodafone's bid from $6 to $9 a share, making it effectively a $62-billion offer. To come up with this much cash, Vodafone had arranged for a $24-billion loan from a syndicate of banks, including BankAmerica Securities, Citibank, Barclays, and Goldman, Sachs & Co. Gent was crawling way out on a limb to grab this deal.

As the executive team entered the Board meeting on Friday afternoon, the presentation they were about to make was like a Harvard Business School case study, with two fundamentally different propositions on the table. One offered a domestic strategy that would merge East and West, making the AirTouch and Bell Atlantic combination the largest wireless company in the United States. This choice focused on consolidation to streamline operations and gain economies of scale. The second strategy was much more ambitious. It promised growth and more growth. Keep building out, Vodafone said, to conquer the wireless world.

The management teams and corporate cultures at Vodafone and Bell Atlantic were so different that it was hard to believe the two organizations were in the same industry. Most of the staff at AirTouch dreaded a return to the Bell culture they thought they had left behind. Bell Atlantic was still basically a local operating company. Many people remembered all the arguments in favor of the spin in 1993. Why, they asked, would AirTouch want to go back to being a Bell company? In Vodafone they saw clear evidence of entrepreneurial values more like the ones they were familiar with in Silicon Valley and San Francisco. In reality, AirTouch's culture was a blend of Bell values and Information Age, entrepreneurial concepts, but the staff identified primarily with the latter perspective.

At the Board meeting, there was, of course, more interest in value for the stockholders than cultural values. Paul Taubman made the presentation for Morgan Stanley. For two weeks Taubman and the corporate development team at AirTouch had engaged in an amicable but heated debate over the relative merits of the Vodafone and Bell Atlantic offers. Morgan Stanley believed that Vodafone's stock price would never hold up. In the long run, Morgan Stanley said, shareholders would be better off with Bell Atlantic. Even after Morgan Stanley's London office expressed confidence in Vodafone's value, Taubman and his team remained concerned. The Air-Touch staff argued that Morgan Stanley's Wall Street analysts just didn't understand the European market. For years Wall Street had failed to give AirTouch full value for its European assets, and this had been a constant source of concern for Ginn and Sarin. But there was no arguing with the market. Morgan Stanley had predicted that, with the announcement of an AirTouch offer, Vodafone's stock would drop substantially. To the contrary, when Wall Street learned of Vodafone's offer, the shares of both Vodafone and AirTouch had gone up. That reaction had answered a lot of questions around AirTouch headquarters.[16]

Vodafone's eleventh-hour decision to increase its bid tipped the scales with the AirTouch directors. The offer represented a substantially better value for shareholders.[17] In their analysis of the two offers, Paul Taubman and his colleagues were professional and tried to remain unbiased. But some of the AirTouch staffers sensed their disappointment. After the presentations, the Board questioned Taubman, Ginn, and Sarin at length, focusing on the competing strategies. Ginn, too, tried to play the situation even-handedly. He wanted the Board to make the decision. The hardest thing for the Board to swallow was the difference between AirTouch's and Vodafone's earnings and share prices. With 50 percent more in operating profits than Vodafone, AirTouch was valued by investors at roughly the same price. Analysts attributed this difference to the sluggish growth of the American cellular market in 1998, where penetration rose by only 5 percent, compared to 10-percent increases in Europe.[18] But still, AirTouch seemed to have a more efficient organization.

Vodafone was also attractive, however, from a human resources point of view. A merger with Bell Atlantic would result in substantial layoffs at AirTouch corporate headquarters as Bell Atlantic consolidated its domestic wireless operations. With Vodafone, however, these layoffs wouldn't be necessary. Vodafone would still need most of the AirTouch personnel to manage its U.S. operations. The idea that this combination would more likely be a merger of equals, with equal representation on the board and an integration of top management, also made a transaction with Vodafone seem less like an end than a beginning. According to Ginn, "There was a lot of feeling that you had two powerful companies here, you brought them together, and culturally they would be better because you brought them together."[19]

✳

Downstairs, "like a lady in waiting," Amy Damianakes could hardly stand the suspense. It would be her job to communicate the final decision to AirTouch's 13,000 employees worldwide. As the planned time to adjourn came and passed, she went upstairs to find that the board members had already left. Ginn's staff was milling around the boardroom. "They were so staid. I didn't know if they were exhausted because they had been working around the clock for days, or because they were in a state of shock." Turning to one staff member, Damianakes asked, "Do we have a deal?"

"Yes, we do," he replied. "It's Vodafone."

In his office, Ginn had just finished giving Chris Gent the news, concluding with an agreement to fly to London so the two executives could announce the decision formally on Monday. The deal, which was enormous, reflected the tremendous accomplishments of these two relatively young organizations and their entrepreneurial executives. Based on the total number of subscribers, AirTouch was already the largest wireless company in the world. The merger with Vodafone created a global powerhouse with a combined market capitalization of nearly $110 billion, making it the third largest publicly traded company in the United Kingdom. The combine had the potential to serve nearly a billion people in 23 countries on four continents.[20] Sam Ginn was pleased, but he was also sad.[21] The vision that his company embodied would now become someone else's dream. As Schumpeter had pointed out long ago, capitalist innovation was both creative and destructive. It generated greater income for the many and made successful entrepreneurs wealthy, but it destroyed existing institutions and forced painful changes on those who were left behind. Sam Ginn was appreciating both sides of that coin as the deal was completed.

On the phone, Ivan Seidenberg told Ginn he understood the situation. He regretted that investors had not supported Bell Atlantic's bid the way they had responded to the news of Vodafone's offer. But he was clearly frustrated.

"These domestic assets still belong together," Ginn said, hoping to keep the door open to a different kind of deal in the United States sometime in the future.

But Seidenberg was noncommittal. He told Ginn that Bell Atlantic would probably terminate the PCS PrimeCo partnership.[22] Ginn saw that as a strategic ploy and wasn't worried. He hoped things could be smoothed over, but first he had to go to London.

TWENTY-THREE

The Meld

Barely 24 hours after the Board's decision, Ginn's driver arrived at their house in Hillsborough. Ginn loaded his bags, and as the car eased down the curving driveway, he caught a commanding view of the San Francisco Bay and of the jets in their final approach to San Francisco International. The dark, waxy leaves of the ceanothus bushes on the hillside reflected the California sunlight. Down the hill and through the black metal gate, the car passed tall eucalyptus trees and then skirted the rich green fairways of the local country club while Ginn organized his thoughts about the days ahead.

He and a small group of his top executives were going to London to take the next step in the merger. With Mohan Gyani and April Walden, the head of AirTouch's investor relations group, he boarded a British Airways red-eye flight for London. He and Gent would hold a press conference on Monday. The formal announcement would launch a long process of global consolidation. In the coming months, AirTouch and Vodafone would need to secure the government approvals necessary to complete the deal and integrate the two organizations – their management, their procedures, their cultures, and their boards. If possible, they would also have to find a way to forge an agreement that would be acceptable to Bell Atlantic because they had yet to establish a national presence in the U.S. market, a critical strategic consideration. Ginn knew that he and Gent also had a partner in Europe – Mannesmann – who would be concerned about the deal.

Beyond these immediate tasks, a larger turning point loomed. The wireless industry was preparing to roll out its third generation of transmission technology. Once again a squabble over standards was taking place, only this time the struggle was shaping up as the United States versus Europe.

The cross-border combination of Vodafone and AirTouch could play a leading role in resolving this dispute, but there was some risk that their new creation could be torn down the middle by the fighting. All of these issues were in Ginn's mind, even though he knew that he now had to step back from day-to-day operations and adjust to a new role as nonexecutive chairman. He was no longer in command.

At the press conference, Ginn broke from British tradition and extemporized on the revolution taking place in telecommunications. "Ten years ago," he said, "you had never heard the name Vodafone. And five years ago you had never heard the name AirTouch." But the new Vodafone AirTouch Plc would be Britain's largest phone company and the third-largest enterprise on the London Stock Exchange.[1] The Vodafone AirTouch brand, he predicted, would quickly become the Coca-Cola of mobile telephony.

With the formal announcement in Britain completed, the AirTouch group still had to return and repeat the exercise in New York. That evening, at Gent's insistence, they headed for Heathrow – accompanied by Gent, Ken Hydon, and others from Vodafone – to take the Concorde flight. Halfway to the airport, Gent realized he had forgotten his luggage. He called his office, and the staff hired a motorcycle to race through traffic to meet the two CEOs just as they checked in for the plane. Moments before they boarded the plane, Gent's cell phone rang. It was Ivan Seidenberg offering his congratulations. "Your shareholders supported you," he said, "and my shareholders didn't support this initiative for me." Brief and seemingly insignificant, the call was actually a clear and important signal: Seidenberg and Bell Atlantic were still interested in cutting a deal for AirTouch's North American assets.[2]

The flight to New York was uneventful but, as the Concorde approached JFK, fierce turbulence and lightning slammed into the jet. Warned away from the airport, the pilot pulled out of his descent and flew to Atlantic City instead. Because there were no customs facilities, the crew had to keep all passengers on board for nearly three hours until the storm let up. Others might have taken the weather as a harbinger of things to come, but Chris Gent and Ken Hydon spent the three hours on the runway making jokes and teaching the Californians to play a card game. "We drank and ate everything on the plane," April Walden said.[3]

✳

The next day, at eight in the morning, the two CEOs appeared before a crowd of Wall Street analysts and reporters, who gave them a standing ovation. It was quite a contrast to the disdain that Ginn and the rest of the

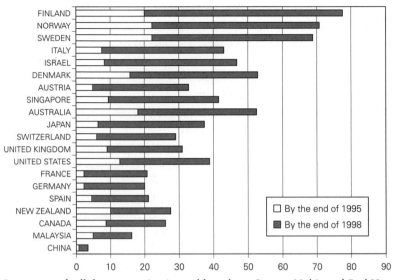

Percentage of cellular penetration in world markets. *Source:* Nokia and *Red Herring* (April 2000).

PacTel team had faced on the roadshow in 1993, when Nextel and the PCS auctions seemed to threaten the viability of their new wireless company. Now AirTouch, Vodafone, and wireless were all in the investor's spotlight. The deal underscored the remarkable growth of the European mobile market following the introduction of GSM in the mid-1990s. As the Europeans were pleased to explain, the standardization of transmission technology had created greater economies of scale in manufacturing equipment and handsets, lowering the costs to consumers. There was, as always, another side to the coin. The weaknesses of Europe's monopoly wireline companies had created greater latent demand for telecommunications services, and wireless had filled this gap.[4] Without a doubt, Europe's firms had responded to this opportunity creatively, encouraging cell-phone use with innovations such as the "calling party pays" plan, and now they were riding high – led by Vodafone.[5]

The publicity around the acquisition also brought attention to the differences between American and British compensation systems. The British press was intensely interested in how much Ginn personally would make off the deal. As Vodafone's share price rocketed upward, climbing 14 percent on news of the agreement, his shares and options stood to net him several hundred million dollars. Others shared in the kind of entrepreneurial profits that always raise the hackles of those in the media and elsewhere who are convinced that equality is more important than innovation. Six other key executives, including Arun and CFO Mohan Gyani, would receive multimillion-dollar returns. In this case, entrepreneurial returns reached deeply into the organization: virtually every one of AirTouch's employees was a shareholder. Together they were eligible for 9 million AirTouch shares worth an estimated $900 million. As the press noted, hundreds of employees who had been with the company since the spin-off would become millionaires as a result of the deal.[6]

Ginn wasn't certain that many of them would stay with the combined company once the deal was complete, but he took great satisfaction in the "bonus" they were now receiving for their efforts in building the enterprise. While the stock-sharing plan had been implemented primarily to motivate the employees, Ginn's attitude toward this sort of compensation system also had a class dimension. He had not forgotten his roots in working-class Alabama. That's why he was so enormously pleased to see $900 million flowing to the people who didn't work on the top floor.

With their work finished in New York, Ginn and Gent flew on to Boston and Minneapolis for more media presentations. Their transportation woes continued. With snow and ice on the ground in Minneapolis, Ginn's limousine from the airport spun out on a freeway off-ramp. Sam looked out through the back window of the car in time to see an out-of-control SUV hurtling toward them. He tried to slide down to the floor to protect himself from the impact, but when the crash came, the trunk folded, smashing the luggage in the rear. The impact drove the back seat into the small of Ginn's back and flattened him onto the floor. Afraid that the car would burst into flames, he and his fellow passenger struggled to get out of the vehicle, but the doors were crimped shut. They waited anxiously for rescue workers to arrive and pry the doors open.[7] The next day, his back extremely sore, Ginn stretched out on the floor to participate in conference calls with investors.

Ginn was not the only one to express relief when the team finally reached San Francisco. At the airport, an AirTouch driver met Chris Gent to take him into the city for an all-hands meeting with AirTouch employees. He was the same driver who had driven Gent to Bohemian Grove earlier in the summer. As he pulled away from the curb, Gent asked him how he felt about the deal. He admitted what many AirTouch employees were thinking: "Thank God it's not Bell Atlantic."[8]

Wall Street, too, was overjoyed. AirTouch brought to the combination a strong focus on service quality and sound engineering. Vodafone added a brash, risk-taking character with a powerful marketing bent and imperial ambitions. Both companies were dedicated to the future of wireless with the kind of intensity that one usually finds among religious missionaries. Neither was distracted by Internet ventures or landline investments; they were deeply committed to the pure-play strategy. What Arun Sarin said about AirTouch was true for Vodafone as well: "We woke up thinking about wireless, went to work and thought about wireless, and at night we dreamed about wireless."

Gent had little desire to impose Vodafone's culture or style of operations on AirTouch, an organization that he frankly admired. His hands-off attitude was true to the history of Vodafone, which had always been a decentralized company that granted foreign subsidiaries a great deal of autonomy. AirTouch employees who watched him perform at the all-hands meeting in San Francisco were impressed. He gave the stage to Ginn to enjoy the outpouring of emotion from employees who expressed their gratitude for his leadership. They were, after all, celebrating their collective accomplishment in more than tripling the company's net worth in six years. Ginn's focus on culture had paid off in terms of shareholder value and employee satisfaction. AirTouch had climbed into the ranks of the top 10 percent of companies in the nation in terms of employee satisfaction by 1998.[9] "People at AirTouch talked these principles, and most of them walked them as well," said one Vodafone manager. For his part, Gent said that Vodafone had a great deal to learn from AirTouch.[10]

Vodafone's values were considerably less well defined. "If you talked to twenty people, you would probably get twenty different perspectives," one Vodafone manager commented. Like most start-ups – and most British companies, for that matter – the firm had focused on signing up subscribers and paid little attention to the softer side of management. But already Gent had started to change that aspect of his business, influenced in part by what he had observed at AirTouch. Shortly after he returned from his visit to Bohemian Grove, for example, he had implemented an employee stock ownership program.[11]

With the acquisition, Gent embraced several other aspects of the Air-Touch approach to culture. He hired Bain & Company to do an integration study that began with an assessment of Vodafone's organizational values. Many people in Gent's company were skeptical. It seemed to be a California thing to do, definitely not very British. They were right, of course, and the Bain report nailed down the differences. It said that AirTouch, still harboring some of its Bell System heritage, took the "ready, aim, aim, fire" approach; Vodafone was more "ready, fire, aim." Watching the two companies from a distance, Craig McCaw agreed with this assessment.

AirTouch was disciplined, he said, and had not fallen prey to the laziness that can affect companies operating in a duopoly. Vodafone, on the other hand, had moved very quickly. "They woke up every morning," McCaw said, "and found some new fight to start."[12]

These two organizational cultures were to a considerable extent shaped by the managerial styles of the entrepreneurs who had created them. Ginn had led AirTouch as a patriarch. He personally identified with the business and the employees and engendered a great deal of personal loyalty. Gent was more distant. He had less personal identification with the business. He was at heart a deal-maker, and this gave him greater flexibility. As a manager, he was adaptable – quick to experiment, and quick to abandon what didn't work and look for something that might improve the business. He epitomized Schumpeter's concept of "creative destruction" and had at Vodafone created a company designed for rapid innovation. In Newbury, decisions were made quickly by a small group of people who worked closely with Gent. Operations, as we have seen, were decentralized, and there was less consultation and consensus building than occurred at AirTouch.[13]

The Vodafone group guiding the integration of the two organizations tried to find a middle ground that would preserve the strengths of each organization. They introduced focus groups, which included 5 percent of all employees. Vodafone developed its own statement of values, modeled after the AirTouch "Compass," and promulgated the new declaration in September 1999. Despite all of these efforts and Vodafone's best intentions, however, there was some residual bad feeling on the American side – if only because it was clear that restructuring would eliminate a number of AirTouch jobs.

Ginn was concerned about his employees. He fought to make sure that AirTouch people were well treated and had fair opportunities to fill open positions and earn promotions.[14] He had a sense of déjà vu, remembering how Clayton Niles had expressed similar feelings when Sam was negotiating his first acquisition, the purchase of Communications Industries. Although his employee stock programs had made millionaires out of hundreds of employees, he was concerned about those who were not wealthy and would need retraining. Setting aside $1 million from his own compensation, he created a Skills Upgrade & Retraining Fund (SURF). The program provided grants worth up to $6,000 each to individual employees to pay for retraining or other educational endeavors that would allow them to pursue new careers or upgrade their skills.[15] In the short term, given the explosive growth of the wireless industry in 1999, few employees who lost their jobs had trouble finding new ones. Those who did could turn to SURF for help.

*

In the immediate aftermath of the AirTouch deal, Chris Gent concentrated on three primary goals: gaining majority control of Vodafone AirTouch's international ventures; expanding the company's global reach; and preparing for the 3G technological transition. Ironically, the first step toward strengthening international control was a move into the back seat of the wireless market in the United States. In February, five weeks after the Air-Touch Board voted to accept the Vodafone offer, Ginn and Arun flew to New York, where they met Chris Gent for the Goldman, Sachs annual industry conference. They also held a hastily arranged meeting between Ginn, Gent, and Ivan Seidenberg at Bell Atlantic's headquarters on the Avenue of the Americas.

It was the first time that Gent had met Seidenberg, so he let the two Americans lead the discussion. Ginn took the initiative and went over the familiar rationale for a merger. He mentioned the need for a North American footprint and the good fit between the Bell Atlantic and AirTouch properties. But as the conversation progressed, the differences between Seidenberg and Ginn resurfaced. Bell Atlantic continued to have sticky issues related to GTE, and Seidenberg said that any deal would have to be structured in a particular way. The biggest sticking point was also unchanged: both sides wanted control.[16]

For Ginn, the meeting was disappointing, "a total waste of time." He was certain that the AirTouch and Bell Atlantic assets belonged together, but Bell Atlantic had ambitions in a lot of different businesses. Its commitment to the traditional wireline side was enormous. Having made the successful spin in 1994, Ginn had become a firm believer in the pure-play wireless enterprise. He thought wireless would never get enough attention if it remained part of the Bell Atlantic structure, and he (like the Department of Justice and Judge Greene) was concerned about the potential for cross-subsidies. In a combined wireless entity, Bell Atlantic would have a 55-percent stake, but it owned 100 percent of its wireline operations. The temptation to use wireless to subsidize the wireline business wouldn't go away and would continue to cause problems with the state and federal governments. Ginn expressed concern that Bell Atlantic's lobbyists would favor the wireline interests in the resulting public debates. He preferred to see both organizations pool their interests in a separate, independent wireless company.[17]

With the potential for a deal growing more remote, Ginn, Sarin, and the rest of their California crew began to look elsewhere to assemble a national network. One option was to acquire PCS companies in the Northeast to expand the firm's footprint. Another was to go after Omnipoint, where George Schmitt was CEO.[18] But then VoiceStream Wireless, which was building a U.S.-based GSM system, acquired Omnipoint as well as Aerial Communications, a Chicago carrier.[19]

Blocked in that direction, Ginn and Gent discussed another path. They could buy the Next Wave entities and carry their projects to completion. Next Wave had bid heavily in the second round (the so-called C Block) of the PCS auctions, which had been organized to benefit smaller, entrepreneurial firms. Next Wave planned to position itself as a wholesaler of wireless capacity.[20] But the business had run into trouble, and the FCC was talking about taking the licenses back. Gent had trouble with this alternative, however, in large part because it would require a substantial investment – somewhere between $10 and $20 billion.[21] He was reluctant to invest that much capital. From his perspective, the combination with Bell Atlantic still looked more attractive because the technologies and the coverage were more compatible.

He was also concerned about timing. They should, Gent said, move quickly. The transition to digital was going more slowly in the United States than in Europe, and competition was intensifying faster than Ginn had predicted. By the spring of 2000, the company was losing customers to digital and to AT&T and Sprint, both of whom offered "no roaming" nationwide service. They were taking some of AirTouch's highest-value business customers. Gent realized that the company had to accelerate the transition to digital, even if it meant reducing margins in the short run. This made a deal with Bell Atlantic even more compelling.

When Warburgs sponsored a conference in London for telecommunications leaders later that year, Gent took the opportunity to sit down with Bell Atlantic's development director, Alex Good. "Look," he said, " I am prepared to look at a deal that recognizes you've got control, but in order for me to make that sacrifice and convince the board of Vodafone Air-Touch, you've got to give us a premium up front in the form of a higher percentage ownership in the joint venture than our contribution of earnings would necessarily warrant."

Good said, "I'll take that message back to Ivan."

Seidenberg agreed to meet with Gent to discuss a possible deal. They got together, but once again Gent found the meeting strange and unproductive.

Chuck Lee, the CEO of GTE and Seidenberg's peer as long as that merger was still pending, was present. But he could say nothing because of the restrictions he was under due to federal rules governing the Bell Atlantic–GTE merger. Gent failed to make any headway, but he had not built Vodafone by giving up easily and he decided to keep this communications channel open.[22] Mohan Gyani tried to break the impasse. He held a series of meetings with Bell Atlantic's CFO Fred Salerno, but he too was disappointed.

Gyani and Gent now realized that they would have to make substantial concessions in order to make the deal work.[23] They presented various options to the combined Vodafone AirTouch Board at a meeting in San Francisco. The Board expressed concern that Vodafone AirTouch risked becoming a passive investment company without operational assets and skills. Nevertheless, the Board agreed that the negotiations should continue. With the Board's support, they flew to New York and camped out for two days in Bell Atlantic's offices. Through the negotiations, the shadow of missed opportunities between AirTouch and Bell Atlantic kept resurfacing for Gyani. Now Gent finally understood why it had taken Air-Touch and Bell Atlantic so long to work out the details for their proposed merger in December. When negotiations stalled, Gent met with Seidenberg, who leaned on his people and made things move along. Nevertheless, according to Gent, "We got to a total brick wall."

Gent told Gyani, "This is clearly not going to work. These guys are just being obdurate. I've got to get back to England." Gent asked Bell Atlantic to arrange for a ride to the airport.

The Bell Atlantic negotiators were shocked. "We're so close to a deal," someone said.

"You're not close at all," Gent replied. "We have an outline and that's it. The minority protections are not resolved, the IPO is not nailed down, nor is the question of how we exit if necessary." He was not bluffing. He clearly intended to leave.

At this point, Seidenberg intervened, promising to have all the issues resolved by the next day. Gent agreed to stay for 24 hours.[24] This time Seidenberg leaned even harder, and his team quickly finished an acceptable deal. When it was announced on September 21, Seidenberg told the press, "This is a logical fit, naturally uniting our U.S. properties and strong management teams, and enhancing the benefits of Bell Atlantic's merger with GTE." For Vodafone, the merger offered an opportunity to maximize the value of its holdings in the United States without having to make substantial additional investments. As one analyst noted, "it allows Vodafone to pursue their global wireless ambitions without having to fight their way into the East Coast market."[25]

The new business, owned 45 percent by Vodafone AirTouch and 55 percent by Bell Atlantic–GTE, would have an estimated value of more than $70 billion. Seven directors would govern the new enterprise: four from Bell Atlantic and three from Vodafone AirTouch. The consolidated company's 30,000 employees would serve 29 million U.S. customers in 49 of the nation's top 50 markets (96 of the top 100 after the completion of GTE's merger with Bell Atlantic).[26] With 28 percent of the U.S. market, the venture would immediately become the largest wireless operator in the country.[27] Under terms of the deal, Vodafone AirTouch also transferred $4.5 billion of debt to the new venture, freeing up capital for Gent to make important moves abroad.[28] The agreement included a guarantee that the new entity would go public within three years, a provision that protected Vodafone in case the relationship with Bell Atlantic became problematic.

It took six months to integrate the two companies. The effort was complicated by the fact that Vodafone was still working on its combination with AirTouch and Bell Atlantic faced similar issues with GTE. For employees in the trenches, the pace of change was head-spinning. But by April, they were able to launch the new company, Verizon Wireless. Following the completion of its merger with GTE, Bell Atlantic adopted Verizon – a combination of the words *veritas* and *horizon* – as its new corporate name. Verizon planned to complete an initial public offering of the wireless company's stock in what was expected to be one of the largest IPOs in history, raising $12 to $15 billion in cash and putting a total market value on the business of between $120 and $150 billion.

The Verizon Wireless deal once more accelerated the race to establish nationwide brands in the United States. Sprint and Nextel had shot ahead in 1994, but both had been slowed by the enormous investments needed to build a nationwide system. Sprint was in a strong position with its PCS license acquisitions, but the build-out of its network took longer than either Sprint or its cable company partners had anticipated. By the spring of 1998, the company was offering service in 157 markets but was covering only about 50 percent of the U.S. population.[29] Nextel had floundered until Craig McCaw agreed to invest more than $1 billion in the company in 1995. McCaw shifted the firm's strategy, going for nationwide coverage but aiming at high-value business users rather than mass-market consumers.[30] AT&T, which had leapt ahead with its One Rate program, had the first truly nationwide single-package service, but the combination of AirTouch and Bell Atlantic's systems suddenly made Verizon the new market leader. Others raced to catch up. Three days after Bell Atlantic unveiled the Verizon name, two other Baby Bells – SBC Communications and BellSouth – announced that they would combine their domestic wireless assets under the brand name Cingular, creating the nation's second-largest wireless company.[31] Suddenly AT&T had dropped two places going down the backstretch, yet another signal that the old world order in telecommunications was changing.

Deregulation, privatization, and now intense competition were transforming global telecommunications. In the large U.S. market, however, neither economies of scale nor the Bell traditions of engineering and operational excellence had given way entirely to new-age entrepreneurs and startups. Indeed, in wireless, the top three players in what was rapidly becoming a national oligopoly all had deep roots in the old Bell System. Verizon, Cingular, and AT&T all had distinguished Bell System credentials. But all three had also been transformed by the technological and economic earthquakes that had unsettled the industry since the Bell breakup. An era of rapid entrepreneurial transformation had seen to that – before giving way to a new era of consolidation. Sam Ginn personified the combination of traditional and new values involved in wireless during both eras.

*

But Ginn was now standing near the sidelines. With U.S. operations under Bell Atlantic's leadership, the good will and sense of teamwork that characterized the beginning of the Vodafone AirTouch combination began to fade. As Seidenberg told reporters, Verizon Wireless would produce $7.4 billion in savings for the companies. Some of those dollars came from layoffs in the Bay Area among Vodafone AirTouch's 1,000 employees.[32] They also came from attrition in the executive ranks, as many of AirTouch's top people started to move on to new ventures. Gent offered some of them positions in the United Kingdom, but most were unwilling to leave California.

One of them was Mohan Gyani.[33] Gent had offered him a position as chief strategist for Vodafone AirTouch, and he had carefully considered the offer. He and his family visited England for the Wimbledon tennis tournament and spent some time in London. But ultimately they decided not to accept Gent's offer. Gyani left the company and shortly was recruited to become chief financial officer of AT&T Wireless. He had barely been at this new job a month when the company's president, Dan Hesse, stunned the business world by announcing he was leaving to join a telecommunications startup. Gyani became president of AT&T Wireless just in time to oversee the company's IPO in April 2000.[34]

While Gent was disappointed to see Gyani go, he still hoped to retain Arun Sarin. A year earlier, the conversation between Vodafone and Air-Touch had broken down over the issue of Sarin's future and the structure of a combined company's management. Gent understood Sarin's value, but his loyalty to the tight inner circle at Vodafone made it difficult for him to make Sarin his chief operating officer. Instead, he gave Sarin the title of CEO for the Asia–Pacific region. But as Gent soon discovered, Sarin brought more to the business than operating ability. His close relationships with AirTouch's former shareholders and the U.S. investment community proved invaluable when Gent and Vodafone AirTouch suddenly got into the biggest corporate fight Europe had ever seen.

TWENTY-FOUR

A Global Powerhouse

Nine years after winning the German D2 license, Mannesmann looked like a textbook case of how to turn around a company. Under the leadership of Werner Dieter and his successor Joachim Funk, Mannesmann had increasingly shifted its assets from the slow-growing steel tubing business into wireless and fixed-line telecommunications in Europe. The company purchased nearly 75 percent of the equity of Austrian fixed-network operator Tele.ring. With Olivetti, it launched a fixed network in Italy and then made a successful bid to acquire Telecom Italia, giving it stakes in the Italian mobile company Omnitel and the fixed-line business Infostrada.[1] It owned 15 percent of the French telecommunications company Cegetel.[2] With the merger of Vodafone and AirTouch, Mannesmann became the second-largest mobile-phone company in Europe.

But Klaus Esser, the heir apparent to succeed Funk as CEO of Mannesmann, was worried. Early in December of 1998, Esser had called Gent to say he wanted to talk about possible cooperation between Mannesmann and Vodafone. This led to a business lunch at the Terence Conran's restaurant in London, a favorite spot for Tony Blair, who took Bill Clinton there when the U.S. President was in the city. Esser and Gent began their lunch on the same day that the AirTouch Board was considering its decision on the Vodafone offer. Esser said to Gent, "When I phoned you up, I had a plan in my mind because I thought that Bell Atlantic and AirTouch were going to come together and then come after me. I thought I ought to seek some kind of defense to that. Now I read it could be you." Esser then asked, "How do you see the future?"

"I think our companies belong together," Gent replied. But he said Vodafone AirTouch was not in a hurry. "I'm quite happy to wait until you're ready. I think we should work together and cooperate. We don't like forcing matters."

At the end of lunch, the two men agreed to wait and see how the Air-Touch bid turned out. Then they said good-bye. Gent, the empire builder, believed that – after the AirTouch combination – a deal with Mannesmann was inevitable. Esser, however, was not interested in being part of a British business empire. He wanted Mannesmann to remain in control of its own destiny, and instead of cooperation he began to think about mounting an aggressive defense against Vodafone. Three days after AirTouch's announcement that it had accepted the Vodafone offer, Esser called Arun Sarin. Given Vodafone's equity interest in E-Plus, he said, Mannesmann believed AirTouch now had a substantial conflict of interest in its relationship with Mannesmann Mobilfunk. In a letter to Ginn, CEO Funk proposed that AirTouch sell its stake in Mannesmann Mobilfunk back to Mannesmann. In any event, he wrote, AirTouch should not participate in meetings of Mannesmann Mobilfunk's supervisory board while the resolution of the merger with Vodafone was pending.[3]

Ginn politely deflected Funk's offer. Vodafone, he wrote, had already announced its intention to sell its E-Plus stake. Until that sale was final and Vodafone had completed its acquisition of AirTouch, Vodafone would receive no proprietary information on the D2 system.[4] In February, Ginn reiterated these points in person. In a Sunday morning meeting with Funk in New York, he expressed his desire that their two companies continue their successful cooperation in Europe. They were partners in the Omnitel wireless operation in Italy and in Mannesmann's long-distance company, Arcor. As long as they continued to work together as partners, Ginn said, Vodafone AirTouch would not launch a bid for Mannesmann.

But shortly after this meeting, cooperation gave way to tension and a threat of conflict. Funk retired, and when Esser took over in May, he made it clear that he no longer viewed Vodafone AirTouch as a partner. Market watchers speculated that Mannesmann would try to buy One 2 One in the United Kingdom and become Vodafone's competitor. Esser, however, surprised everyone.

Ginn was at the annual planning meeting for Vodafone AirTouch in England in October when he got the news. He and Gent had been scheduled to fly to Dusseldorf to help celebrate the ten-year anniversary of the Mannesmann Mobilfunk partnership. But on the day before they were scheduled to leave, Esser announced he was negotiating with Hutchison Whampoa to buy Orange.[5] Deutsche Telekom had acquired One 2 One, closing off that door to the British market, so Mannesman made a highball offer for Orange. Esser was challenging Gent in his own backyard, and he abruptly cancelled the trip to Germany and the joint celebration. Mannesmann's move diminished the opportunities for Vodafone to acquire control in Germany and Italy, and it made Gent's dominant partner his leading competitor.

But to both Gent and Sam Ginn, Esser looked vulnerable. Mannesmann's $34.4-billion bid for Orange was, they thought, extravagant. Esser was paying almost twice as much per subscriber as Deutsche Telekom had paid for One 2 One only two months earlier. Mannesmann issued new shares to engineer the deal, but shareholders responded by promptly dumping the company's stock, sending it down 10 percent. Esser was in a bind, trapped by a market that had sharply reduced his degrees of freedom.[6]

Chris Gent decided to move quickly and forcefully to take advantage of the trap. In November, Vodafone AirTouch offered to merge with Mannesmann, but Esser rebuffed Gent's $105-billion stock-swap offer, calling it "wholly inadequate."[7] Given Esser's refusal to discuss a proposal, Gent went to his Board for approval to submit an offer directly to Mannesmann's shareholders. This was a risky move. No hostile takeover had ever been successful in Germany, whose laws handicap foreign companies mounting unsolicited attacks. Unlike the United States and Britain, Germany did not have an open market for corporate control. This was one aspect of globalization that ran counter to the cultures and political economies of Germany as well as Japan – nations that had enjoyed tremendous economic success in the latter stages of the Second Industrial Revolution. Now, however, the Information Age was knocking on the door in Germany in the person of Chris Gent, a distinctly Third Industrial Revolution entrepreneur. When Gent first knocked, the door was tightly closed. More than half the seats on Mannesmann's supervisory board were held by trade-union representatives who could be expected to oppose any Vodafone offer, fearing that Mannesmann's industrial companies would be liquidated. Nevertheless, Gent was determined to press forward. As he told the Board on November 18, "We will be damned if we don't go. If we fail, we will be damned as well." Then, in words that would have been applauded by every innovator since the dawn of capitalism, he said, "I would rather be damned for trying something than for sitting on my hands."[8]

Gent's support was not unanimous. Michael Boskin, an AirTouch board member who had joined the combined board, agreed with Gent. "This guy stiffed us," he said. Vodafone AirTouch had every reason to go after Mannesmann.[9] However, the investment bankers were less certain about that. Usually bankers push for a deal, but in this case Warburgs and Goldman, Sachs were nervous. For one thing, Goldman, Sachs had represented Mannesmann in an earlier transaction. The bankers feared that German government officials would intervene. National pride was at stake, and rumblings were heard throughout the country. Vodafone AirTouch had to push the bankers to pursue the deal, and Gent pushed hard.[10]

The offer started a telling battle for the hearts, minds, and pocketbooks of Mannesmann's shareholders. Win or lose, the deal would have enormous consequences for them, for Vodafone, and for the German economy. Gent believed that failure would cost him his job. Esser believed the same thing. Investors worried that if a deal did not go through, the D2 partnership would be poisoned, killing their golden goose. That gave Gent some leverage, but at first Mannesmann and German tradition had the upper hand. When Vodafone AirTouch's hostile offer was announced on November 15, Mannesmann's shares rose 10 percent while Vodafone's dropped nearly 20 percent.[11] Nevertheless, Mannesmann's market value did not climb to anything near the Vodafone offer. Investors clearly expected the offer to fail. Esser could expect to have the support of German shareholders, and that meant he needed the votes of only 12 percent of the shareholders outside Germany to stop Vodafone. Given the loyalty of some shareholders – particularly Prudential, Schroders, and Standard Life, as well as the AFL-CIO in the United States (which owned 13 percent of Mannesmann) – Esser seemed to be holding a pat hand.[12] In addition, Hutchison Whampoa, which owned 10 percent of Mannesmann's shares, backed Esser in the struggle. Ominously, however, they conditioned their pledge, saying they would not block the deal if more than 50 percent of the company's shareholders approved the acquisition.[13]

When the stock of Vodafone AirTouch started to rise, the value of the Mannesmann offer climbed to $155 billion by January 2000. Under the terms of the proposal, Mannesmann shareholders would receive 47.2 percent of the combined company. Klaus Esser claimed this was woefully inadequate. Then, he made a tactical error by asserting that his shareholders deserved something closer to a 58 percent stake.[14] He had overreached himself: his demands were so high that it began to appear he was defending his control, not his shareholders. Gent asked why, if Mannesmann planned to issue new shares to finance the Orange acquisition at 155 euros a share, Esser now thought his company was worth twice that amount?[15] This was a question for which Esser could provide no believable answer. He had crawled far out on a limb that started to give way.

The Vodafone AirTouch takeover bid posed a crucial question for Germany – the strongest economy in Europe – and for the rest of the European community. Would Germany play a leading role in the Information Age or would it give way to more flexible global players? The bid also marked a crucial transition in the globalization of the wireless industry. The market for corporate control in Europe had been gradually opening up for some time as nationalistic barriers fell. The era of national champions and PTTs was ending, and the consolidation movement in wireless was pushing the industry toward global oligopoly. Companies that could achieve formidable economies of scale and maintain international footprints were just around the corner.

When some writers suggested that Germany would fight the takeover of Mannesmann's network to protect the interests of the state, Gent scoffed, "This isn't going to be decided by some Fortress Germany."[16] He was right. Although the Vodafone AirTouch bid outraged German Chancellor Gerhard Schroeder – who attacked it as a violation of Germany's tradition of consensus management and suggested that it might lead to the dismantling of a historic and vital German firm – the government decided not to intervene officially.[17] To his credit, Esser also resisted governmental involvement, even if it might have helped his cause. He recognized that it would ultimately be counterproductive for the European economy if the German government stepped in.[18]

As the battle for Mannesmann continued, the convergence strategy again became an issue. Esser argued that Mannesmann's investments in both fixed and wireless service would enable the company to bundle services, a better option than Vodafone AirTouch's pure-play strategy.[19] At a press conference at the Savoy Hotel in London, Esser sought to persuade a small army of analysts and journalists that Mannesmann's approach was superior. He explained that Mannesmann, through its Arcor subsidiary, had invested heavily in fixed-wire systems in Germany, Austria, and Italy. This investment, Esser argued, would give the company a leg up.

Esser scrambled to promote his strategy, but his analyses were considerably short of overwhelming. The convergence theme rang a little hollow. Sitting beside Esser was Mannesmann's new golden boy, Hans Snook, the CEO of Orange. Snook, like Sam Ginn and Chris Gent, had long been an ardent advocate of the pure-play strategy for wireless.[20] Esser also tried to drive home a second point, claiming that Mannesmann's assets were valuable because the firm controlled more of them, in contrast to Vodafone AirTouch, which owned only a minority position in many of its overseas operations (including Verizon in the United States). This argument also failed to carry much weight. Esser was looking more and more like a man who just didn't want to lose his job, and as Gent pointed out, it was better to own 45 percent of the largest coast-to-coast wireless company in the United States than 100 percent of a regional player.

Gent, who was a professional optimist, remained hopeful. He was betting heavily on the fact that 60 percent of Mannesmann's shareholders were not German. After Mannesmann acquired Orange, that percentage increased to 70.[21] Vodafone AirTouch had offered those shareholders an attractive deal that Gent thought they would find hard to refuse. He, Hydon, and Horn-Smith pitched their case to European investors. Arun pounded away at Wall Street and the American markets. They could all sense that, as investors analyzed Mannesmann's arguments in January, their opinions were shifting away from Esser and toward Gent. Money talked and, after three months, investors found it prudent to listen. Vodafone AirTouch's offer effectively doubled the value of Mannesmann's shares between November 2000 and the first of February 2001, and an increasing number of arbitrageurs, anxious to profit from the difference between the offer and the price of Mannesmann, began to buy the German company's stock. They, of course, favored a deal.[22]

Meanwhile, Vodafone AirTouch's shares went up by 23 percent.[23] By late January, Gent, Sarin, and the rest of the Vodafone AirTouch leadership began to sense victory. Several key analysts recommended that shareholders accept the offer.[24] As the pressure on Mannesmann mounted, Esser could see that his support was ebbing. He finally agreed to a deal.

On February 3, Vodafone and Mannesmann announced a friendly agreement to combine the two companies. The combination was valued at more than 181.4 billion euros ($179.66 billion) and was considered the world's largest-ever corporate marriage.[25] The agreement pushed Vodafone AirTouch into an even more commanding lead in the worldwide wireless race. It created the largest company in Europe, with a market capitalization of about £200 billion.[26] As one analyst pointed out, a key factor for investors had probably been Vodafone's global strategy. Although Mannesmann had argued that it, too, planned to become a global player, Vodafone was already there.[27]

With the negotiations complete, Vodafone AirTouch had to sell Orange, and France Telecom came in with the best offer. This $37.4-billion sale was announced at the end of May 2000.[28] Still on the move, Vodafone AirTouch announced that it would use the cash from the transaction to fuel further global expansion, particularly in Asia.[29] You had to respect that announcement, coming as it did from a firm that within two years had increased its customer base from 5.8 million to 42 million.[30] Its reach had widened from 13 to 25 nations on five continents, with the potential to serve more than half a billion people.[31] All of this growth was still being managed from a courtyard office off the High Street in Newbury, where Vodafone had launched its business sixteen years earlier. When visitors came to call on the CEO, they still entered a reception area that had more in common with a dentist's office than the headquarters of one of the world's largest corporations.

In the early summer of 2000, changes were underway, including the construction of new corporate headquarters on the edge of town. But the firm's strategy was still entrepreneurial and still focused securely on growth. Although a slowdown began in telecommunications later that year, Gent steamed ahead with further acquisitions for the firm's global empire. Administrative consolidation aimed at improving efficiency still seemed to be on hold, and Gent now lost Arun Sarin, an executive who might have streamlined operations as he had at AirTouch. But Gent still didn't want a chief operating officer and Sarin wouldn't stay in a lesser capacity. Courted by a number of businesses, he chose to become the CEO of InfoSpace, a wireless industry startup, in April. Disappointed, Gent made a magnanimous offer that also kept his options open: he persuaded Sarin to remain on Vodafone AirTouch's Board.[32]

Vodafone's primary goal was expansion, but the firm started to make some important changes in the way its operations were managed. Among other things, it addressed the issue of compensation. Gent had already adopted the California-style approach that had worked for AirTouch by encouraging employee stock ownership. But as he knew, Vodafone's executives were undercompensated given the scale and scope of the company they had created. British firms had disdained the extravagant levels of CEO pay characteristic of multinational firms in the United States. The rich rewards that AirTouch executives received from stock options following the sale of their company still rubbed many Brits the wrong way – though they were generally polite about it. Gent, however, now rightfully recognized that he and his top executives had built up a major world-class firm, and he decided they were all entitled to a bonus after tripling the value of the company's stock in two years.[33]

Sam Ginn, who was still chairman of Vodafone AirTouch, felt it was a bad move. He had no problem with bringing the compensation for Gent and Vodafone's other top executives in line with that of other leading global companies. He favored the kind of reward-for-performance structures he had employed at AirTouch. But he felt it set a dangerous precedent to pay a bonus on the completion of a deal. Executives should be rewarded, he said, in line with shareholders, as the deal increased the value of the company over time. This put Ginn in an uncomfortable position. He had received substantially more than $100 million in shares and options when AirTouch was sold, and it is hardly surprising that Vodafone's British executives felt betrayed. In the end they had little to complain about because the Vodafone AirTouch Board overrode Ginn and approved a £10-million award to Gent, half to be paid in cash and the other half in stock options. The Board also made £5-million awards to several other executives.[34]

When news of the bonuses hit the British press, a number of investors and analysts reacted angrily. Many in Britain – where the shadows of state-owned enterprise and a full-blown welfare system still lingered – believed that no one was entitled to earn more than a million pounds a year. Britain's National Association of Pension Funds called on its members to vote against the bonuses at Vodafone's annual meeting in July. Echoing Sam Ginn's concerns, *The Economist* criticized the move: "Shareholders are right to be unhappy, for the whole point of big bonuses is to align the interests of shareholders and managers. This does not; it rewards Mr. Gent for a transaction, not for creating and sustaining value."[35] But *The Economist* went on to acknowledge that, among American fund managers, the move hardly raised an eyebrow.

Ignoring the uproar, Chris Gent (who no longer had political ambitions) stayed the course that had brought him and his firm to the top tier in the global telecommunications industry. In Europe, Vodafone already had a controlling interest in the Spanish AirTel.[36] As a result of the Mannesmann acquisition, Vodafone AirTouch also controlled the Italian wireless company Omnitel.[37] Gent purchased a 25-percent stake in a Swiss wireless operator, Swisscom Mobile, in the fall of 2000, and then set out to consolidate his coverage in the United Kingdom.[38] In December, Ireland's Eircom agreed to sell its mobile unit to Vodafone in a stock deal valued at $4.09 billion.[39]

Asia also received Gent's attention and resources. He told the *Wall Street Journal* in November that "We do expect to make acquisitions in the next six months."[40] The issue of foreign ownership in Asia posed some difficult problems, but Gent had a briefcase full of solutions. Only a month earlier, Vodafone had scored a major coup: beating out AT&T, Deutsche Telekom, and NTT DoCoMo by entering into a strategic alliance with China Mobile. Under the terms of this agreement, for $2.5 billion Vodafone acquired a toehold stake (2.18 percent) in what was already the second-largest mobile market in the world, with 60 million customers.[41] An acquisition in Japan followed the Chinese deal. Vodafone bought an additional 15-percent stake in Japan Telecom for £1.5 billion, increasing its ownership in Japan Telecom's J-Phone wireless unit from 26 to 34 percent.[42]

The pace of these acquisitions was staggering, especially in a falling stock market, but Gent was not through. His goal was majority control wherever possible. Two months after increasing his stake in J-Phone, Gent was back with another deal, this time to buy AT&T's 10-percent stake in Japan Telecom for $1.35 billion. Vodafone beat out British Telecom, which had been hoping to close a deal with AT&T but could not come up with a compelling cash offer. This latest move increased Vodafone's ownership in Japan Telecom, and Gent still wasn't finished.[43] In May 2001, Vodafone announced that it would pay £3.7 billion for British Telecom's 20-percent stake in Japan Telecom and other stakes in Japan Telecom's mobile-phone subsidiaries.[44] In less than six months, Vodafone had jacked up its share of Japan Telecom to 45 percent.

Gent's deals highlighted the enormous transition taking place in global telecommunications. AT&T and British Telecom, giants through most of the twentieth century, had bet big on convergence. Yet as telecommunications stocks imploded, particularly in the long-distance and network equipment markets, Vodafone's ability to keep buying and expanding demonstrated the staying power of wireless telephony, the continued prospects for wireless growth, and the continued relevance of the pure-play strategy to entrepreneurship in this industry.[45] As if to underscore this point, British Telecom found itself struggling to convince investors to buy more of its shares in May 2001. Meanwhile, Vodafone quickly stepped into the market and raised $5 billion in a mammoth, record-setting share offering that sold out in just ten hours – despite the fact that its share price had fallen nearly 50 percent in the previous twelve months.[46] The embattled CEO of British Telecom, who had depicted BT's interests in Japan as among the most lustrous of the company's jewels, could only concede that at this point debt reduction had to be his first priority.

*

Despite his focus on global expansion, Gent did not ignore the industry's technical evolution and, in particular, the transition to 3G (third-generation) technology. He focused on two critical elements: (i) taking a commanding lead in the acquisition of 3G spectrum around the world and (ii) positioning Vodafone as the portal of choice to the wireless Web.

Fortunately for Gent and Vodafone, one of the major roadblocks to 3G development had been removed at about the time that Vodafone and Air-Touch were winning approval for their combination. The turning point in the war over 3G standards had come in March 1999, when Ericsson stunned the wireless community by announcing that it would buy Qualcomm's infrastructure division. To see Sven-Christer Nilsson, Ericsson's CEO, smiling as he stood together with Irwin Jacobs seemed as strange to one writer as if Bill Clinton and Special Prosecutor Ken Starr had agreed to "shake hands and declare a budding friendship."[47] Sam Ginn and Craig Farrill could take some credit for this astounding agreement, but the primary responsibility rested with the U.S. government and the pressure it put on China to open its markets to American telecommunications technology. That pressure had focused particularly on Qualcomm and CDMA. The Chinese were considering CDMA for the country's second national wireless system. But they apparently held off making a decision as a bargaining chip in their effort to gain U.S. support for China's entrance into the World Trade Organization.[48] Realizing that Qualcomm's entry into China would provide a giant boost to CDMA's worldwide market share, Ericsson knew that it could no longer afford to continue the war.

The agreement between Ericsson and Qualcomm included a commitment to work together on a worldwide standard for 3G technology. What emerged was a process of reconciliation that allowed time for GSM and CDMA users to migrate. In early 2000, the International Telecommunications Union issued a landmark set of five variations of a common standard for 3G, all based on CDMA.[49]

With the standards war winding down, Vodafone and other cellular companies had to deal with the issues related to designing and financing their 3G networks. The financing produced some severe strategic headaches. The new 3G cell phones were technically attractive. They would provide a direct link to the World Wide Web, allowing customers to remain not only in voice contact anytime, anywhere, but also to have access to images and information whenever and wherever they were needed. But the move to 3G posed business problems that in comparison made the transition from analog to digital seem hardly more difficult than rewiring a light fixture.

To get to 3G, operators would have to convert to broadband packet transmission as well as Internet protocol (IP) packet switching. This would demand enormous capital, much more than it took to switch from analog to digital. Those funds would have to be raised at a time when firms were continuing to expand the capabilities of the second-generation technology, a development that made the transition and the heavy investment seem less urgent. Because most operators had already converted to digital switching in the mid-1990s, they had already paid off a significant portion of their investment in the transition. With the 3G transition, according to Craig Farrill, "We are talking about modifying every network that has been built worldwide."[50]

The transition to 3G resulted in a new battle for the pockets and purses of customers. In the past, cellular operators provided the medium for service; customers created the content as they spoke. As phones became portals for a vast array of information services, network operators had to decide what role they would play in creating the interfaces and generating content. Chris Gent did not want to make the same mistake that AT&T had made in letting America Online create a vast new business on its platform. "We want to avoid that mistake in Europe," Gent said.[51]

He had already launched Vodafone's portal strategy as a flanking measure during the battle with Mannesmann. Concerned that an alliance between Mannesmann and the French company Vivendi would give Mannesmann a controlling interest in the French mobile company Cegetel, Gent went behind enemy lines to have his own conversation with Vivendi. The result was another Vodafone bombshell. At the end of January, Vivendi announced that it would join forces with Vodafone, not Mannesmann, to develop a mobile Web portal to distribute content and services to Vodafone's customers. The new business would be known as Vizzavi.[52]

By the spring of 2000, Gent was telling reporters that he believed Vodafone would eventually become the globe's biggest Internet company. To Americans this seemed like an extraordinary ambition, particularly after AOL's acquisition of Time-Warner. But cell-phone penetration in Europe and Asia exceeded that of personal computers, and as Gent pointed out, many of these customers would encounter the Internet for the first time through their mobile phones.[53] In April 2001, Vodafone moved closer to its goal when the company placed a call over its first 3G network – an Ericsson system using a Panasonic phone.

✳

To cash in on this new strategy, Vodafone needed to secure spectrum, and that was another costly undertaking. The 3G auctions began first in Europe, and those in the United Kingdom were concluded in April 2000. Bidders there offered more than £22 billion ($35 billion) for the five licenses that would enable them to operate on the industry's technological cutting edge. Some questioned whether the price was worth it. Vodafone's license cost nearly £6 billion, and analysts estimated it would cost another

2–5 billion pounds to build the network. As one writer put it, "No wonder investors are nervous. The mobile future is uncertain. It is impossible to prove that the license winners have overpaid. But it is hard to see how they can earn a return on this investment that exceeds their cost of capital and creates value for shareholders."[54] As auctions continued around Europe and as the price for these licenses remained high, investor anxiety increased, especially when it became clear that firms in wireline telecommunications, old and new, had massively overinvested in fixed fiber-optic capacity. The Bank of England warned that telecom companies were taking on so much debt that it could lead to a worldwide financial crisis.[55]

As Europe dashed ahead with 3G, the United States was once again dragging its governmental feet, moving at a lugubrious pace. Although the FCC began its auctions in December 2000, the government didn't plan to have all its 3G licenses auctioned until September 2002. By that time, the Europeans anticipated that their 3G networks would be streaming audio and video to millions of their customers.

They would, that is, if the companies could afford the networks and earn the cost of their capital. That was threatening to be a major problem in both Europe and the United States. In America, for instance, the U.S. analysts anticipated that Verizon might have to spend as much as $15 billion to secure the 3G licenses it would need. This was not good news when the equity markets were turning sour on the high-tech sector. Telecom and other technology stocks peaked in March 2000 and then began a precipitous drop.[56] Vodafone's shares, as we have seen, were not immune to this general collapse. But investors still were prepared to finance Gent's global plans and, in the spring of 2001, Vodafone went to the London stock market and raised another $5 billion in Europe's largest-ever one-day equity sale.[57]

By that time, Vodafone had achieved a commanding lead in the wireless business. With 80 million global customers, it was by a considerable margin the largest telecommunications company in the world. AT&T, Cingular, Sprint, Nextel, NTT DoCoMo, France Telecom–Orange, Deutsche Telekom, and others battled Vodafone in regional markets, but none of them had an equivalent global reach. In a service industry once dominated by huge regulated or state-run institutions, the backroom offshoot of a British defense contractor had thrived on a new competitive order and in just sixteen years found daylight at the top of global telecommunications.

While Chris Gent was pressing forward with expansion, Sam Ginn was stepping back from active involvement in the business he had done so much to develop. Gent had worried that Sam would not be able to give up that important part of his life. But Ginn had seen too many CEOs try to remain in control as chairmen after a merger or a regular succession. In almost every case, he knew, it damaged the organization. In the wake of the Mannesmann settlement, he realized that European issues would dominate the company's immediate business, and he knew that Gent and his top executives were on top of those activities. Then, too, his stand on the bonus issue wore thin his relationships with the British executives. By the late spring of 2000, he was ready to retire and move on to something new. He didn't know what it would be, and he suspected it would not be another "Alexander Graham Bell opportunity." But he knew there would be opportunities for an entrepreneur as the Information Age continued to unfold.

As Ginn was retiring, the twin processes of creative destruction and creative transformation were continuing to roil the wireless world. Even the name AirTouch disappeared in the process. In Europe, AirTouch had never used its own brand, and in the United States it was supplanted by Verizon. Gent proposed to the company's shareholders in July that they revert to the name Vodafone Group Plc. When the shareholders approved his proposal, AirTouch joined the ranks of the hundreds of thousands of corporate organizations that have come and gone in the long waves of innovation that have been the central feature of capitalism's expansion around the world.

TWENTY-FIVE

The Wireless World

In September 2000, when the Summer Olympics began, athletes streamed into the stadium in Sydney, Australia, carrying flags from 200 different nations. From the stands, 110,000 fans cheered, applauded, and waved as Australian Olympian Cathy Freeman, a sprinter of Aboriginal origins, carried the Olympic torch up four sets of white stairs and ignited the cauldron in a burst of flames. Sixteen years after the Summer Games in Los Angeles, Carl Lewis was no longer among the athletes, and Sam Ginn was in California, watching the events on TV and reflecting on a changing world. The "brick" had long since been retired to the Motorola Museum in Schaumberg, Illinois.

As Ginn and others knew, however, the revolution that had begun in 1984 pervaded the Olympic stadium in September 2000. While cameras captured the pageantry of the moment, spectators in the stands and athletes on the field placed an unprecedented 125,000 calls from their cell phones, conversations that equaled nearly five months of talk.[1] What Sam and others had seen as an epiphany in Los Angeles had by 2000 become a commonplace of everyday life. "Leaky" coaxial cable strung throughout the Olympic Stadium made the entire structure a virtual antenna for mobile phones. Homebush Bay, the site of the majority of the activities, was a "sea of glass" because of the fiber-optic cables underneath the walkways in Olympic Park. Over this grid, images and information sped to broadcast facilities and television sets around the world.

These cables, transmitters, electronic switches, and antennae were a vital part of the technological infrastructure of the Third Industrial Revolution.[2] Dramatic innovations in solid-state physics and signaling technology were crucial to this revolution, but as the history of wireless makes clear, the

front across which innovation was taking place was exceptionally broad –
much broader than technology alone. Changes in technology were important, but so too were changes in the political context and a number of
decisive transformations that took place in business practices. This was
especially true during the early, entrepreneurial era of wireless expansion,
when governments as well as business executives in firms large and small
were struggling with unfamiliar situations. It was no more comfortable
for a regulator or legislator than it was for a business executive to be on
the steep part of the learning curve. Yet that's where most of them spent
most of their time during the 1980s and early 1990s.

There is much in the early history of wireless that can best be understood in vintage Schumpeterian terms. Business pioneers like Sam Ginn,
Craig McCaw, and Chris Gent all had personal visions of what could be
accomplished with the new technology and how it should be done. Mc-
Caw was early to the competition at a time when the risks were very high
and the future murky, at best. At the core of his strategy were an appreciation of the value of spectrum and a fine understanding of the present and
future demand for this new form of communication. He underestimated
the market. Everyone did. But he saw that – outside the large metropolitan
areas, in rural and small-town America – wireless would have an appeal
that was being ignored. He correctly divined how the governmentally imposed duopoly structure of the industry in the United States would evolve.
He stuck to his pure-play strategy after convergence became the rage and
larger firms were investing billions to ensure that wireless, wireline communications, cable, and the Internet could all be offered through one-stop
shopping at their firm.[3] McCaw relentlessly extended his network, pressing hard on his supply of funds until finally, unable to raise new capital,
he was forced to sell his corporate creation to AT&T. In a manner that
would have pleased Schumpeter, McCaw loomed larger than the organizations he created.

Chris Gent made his imprint in the second, consolidation phase of the
industry's evolution and continues to play a large role in shaping the industry today. He sought the business counterpart of the nineteenth-century
British empire, with the important difference that his corporate combine
reached into developed as well as underdeveloped countries. His ambitions
were global and his innovations more structural than operational. Were
they as important as the new technology? Yes, we believe they were. They
reshaped, for instance, the response of the largest, most successful economy in Europe to the rise of the global market for corporate control. When
Vodafone, Gent's company, was able to buy Mannesmann, he cracked

Germany's nationalistic resistance to this new aspect of globalization. As a result, German companies, both hunter and hunted, began to leave behind their Second Industrial Revolution heritage of national protectionism and bring their substantial organizational capabilities to bear around the world. It was not long, in fact, before Deutsche Telekom – which caused a congressional flurry by acquiring VoiceStream Communications – performed the same function in the United States with the same result, a surrender of nationalism to a new global standard.

Whether or not Vodafone is able to weather the difficulties created by a depressed market for telecom securities, and whether or not the company is able to develop an organization as effective in operations as it is in mergers and acquisitions, Gent and Vodafone have firmly pointed the way toward the industrial structure that is already becoming dominant in wireless and other high-tech industries. Global oligopoly is upon us in many sectors of the world's industrial economy and is rapidly emerging in others. One of the unanticipated consequences of relatively free trade and international competition, global concentration will bring to industries like wireless the advantages of interconnection and economies of scale. It will also raise the questions that have been asked for a century or more about the market and political power of oligopolists and the effect this structure has upon innovation – only now these questions will be raised about global, not national, entities.

Sam Ginn was a different type of entrepreneur and, more so than Gent and McCaw, he straddled the first two phases of wireless development. Unlike his counterparts, Ginn built a highly successful career in a monopoly enterprise, the Bell System. AT&T was for many decades the largest private corporation in the world, and it was a marvelously intricate bureaucratic organization. Pervasively regulated and periodically subjected to antitrust suits and investigations, the System nevertheless weathered all of its political crises and developed what most observers acknowledged was the most efficient and innovative telecommunications network in the world. Ginn thrived in this environment by mastering the rules, by becoming a talented manager of people and capital, and by demonstrating again and again that he could cope with crises successfully in the Bell way. He was a Bell-head to the core. While he had more skill as a salesman than most of the engineers, he revered the network and knew how to keep it humming. He admired the way the System trained and evaluated its managers and executives, and he stayed on the Bell fast track through this phase of his business career.

Not until he neared the pinnacle of the System did he get a taste of entrepreneurship and decide that he liked the excitement of what he saw as an "Alexander Graham Bell opportunity" in wireless. He had been introduced to the concept of entrepreneurship some years before in an executive course at Stanford University. But not until he took charge of the diversification program of Pacific Telesis Group, one of the Baby Bells, did he begin to understand the economic, political, cultural, and organizational problems that entrepreneurs regularly confront. When he opted to spin off the wireless enterprise from Pacific Telesis and give up his job as CEO of a Fortune 50 company, he had made two decisions that Schumpeter would have applauded.

Ginn turned out to be a third type of entrepreneur: one who kept a foot planted in the Bell way of doing things while creating a fast-moving, innovative global business. From the System, he took a respect for operational excellence and blended it with the cost-conscious, customer-centered values of a young wireless firm. From the System, he inherited a deep commitment to good engineering and managerial development. He combined this with a decentralized structure that would have been an anathema to any well-socialized Bell-head. At AirTouch, he encouraged the development of an organizational culture that stressed individual responsibility, creativity, flexibility, and a global outlook on business and politics. Instead of the sort of control the System had developed – internally and to an unusual degree externally as well – AirTouch engaged in a variety of corporate relationships tailored to fit conditions in each nation, state, and locality in which it was introducing wireless services. AirTouch was a nonunion firm that managed to avoid the kind of adversarial labor–management relations that had characterized most of the developed capitalist economies in the Second Industrial Revolution.

Because Ginn was a bureaucratic entrepreneur who was a skillful builder of organizational capabilities, he doesn't fit the Schumpeterian mold as well as McCaw. Ginn succeeded in building a new and important organization that had a distinctive culture and style of operations. Above all, it was an efficient, innovative organization with substantial technical, managerial, and marketing capabilities. It was also a distinctly Third Industrial Revolution organization, built with highly varied global partnerships that patched together capabilities from a variety of organizations and national cultures. Ginn and his fellow executives had to learn how to do this. There was very little in their Second Industrial Revolution experiences that was particularly useful when they had to become global organization builders.

The early failures Ginn experienced in China and the fumbling manner in which he and his colleagues approached that task give you a good idea of just how steep their learning curve was at first.

AirTouch was a unique company, but it shared a common heritage with other firms that now play a leading role in the wireless industry in the United States and around the world, including Verizon, BellSouth, SBC, and AT&T. Many of the problems that Sam Ginn and his fellow executives faced confront managers at these firms as well. Thus, the Air-Touch story reflects one of the dominant themes in telecommunications. Around the world – as former public telephone and telegraph companies have privatized and expanded beyond their national boundaries – former bureaucratic managers at Deutsche Telekom, NTT, British Telecommunications, France Telecom, and other companies are struggling to become entrepreneurs lest the forces of creative destruction crush their organizations.

The neo-Schumpeterian concepts of Richard Nelson and other scholars working in the discipline of evolutionary economics capture some of the salient aspects of these creative transformations and, in particular, of Ginn's entrepreneurial style. As Nelson and colleagues have pointed out, institutional factors shaped decisively by history play a significant role in guiding the process of economic change. Certainly the Bell tradition was and still is a factor of enormous importance to wireless in this country and in many others as well. By using and adapting that culture and operational tradition to a global Information Age industry, Ginn, Arun Sarin, and the rest of the executives connected with AirTouch were able to create an unusually successful business that passed the test of competition when margins in wireless narrowed and national brands began to be important marketing tools.[4] Their accomplishment reflected the essence of entrepreneurship in an organizational context shaped in part by a history particular to the leading system in U.S. telecommunications.

Even more of Ginn's experience in building AirTouch and in guiding it into the industry's consolidation phase can be explained by reference to the synthesis of Alfred D. Chandler, the world's leading historian of modern, capitalist enterprise. In Chandler's view, economies of scale and scope have been of overwhelming importance to successful industrial enterprises in all of the developed economies of the twentieth century.[5] In wireless during the consolidation phase, scale became a primary factor guiding outcomes in both national and international markets. Only by making tremendous capital investments – to come up to scale and to have a national footprint and later an international presence – could any of the firms succeed. Hence AirTouch was folded into Vodafone and the U.S. operations were later brought into Verizon.

*

What the Chandler synthesis fails to help us understand is the overwhelming importance of the political context to the technical and economic developments in wireless. From the very beginning of wireless transmission early in the twentieth century, political institutions and individuals had a dramatic impact on the industry and its patterns of innovation. More often than not, the political system from Marconi to the present day slowed the process of "creative destruction," allowing interests threatened by innovation to block or at least slow down change. Sam Ginn's experiences with the California Public Utilities Commission provide a casebook study of regulatory contributions to inertia. Like a powerful gyroscope, the CPUC exerted pressure that in normal conditions held the entities it regulated on course. This pattern of organizational behavior was intrinsic to regulatory systems, which were created in the first instance to protect the public in the short term – not to promote innovation in the long term.[6] As the interest-group environment became ever more complex and as competitors learned how to manipulate the system, the dynamics of regulatory evolution made the agencies even more conservative about change. The FCC's performance in the 1980s and 1990s was similar to that of the CPUC. If anything, the federal agency had a more negative impact on the rate of technological and organizational progress because it had greater reach than any state commission.

Two forces, one economic and the other political, finally broke the hold of the regulatory systems and opened the way for changes in technologies, business organizations, and corporate strategies. One was the general movement away from government control and toward the market, a global phenomenon with deep ideological and economic roots that extended back into the 1950s, 1960s, and 1970s.[7] The other was the promise of entrepreneurial profits – the payoff from an "Alexander Graham Bell opportunity," for instance – and the relentless drive that these decisive innovations can produce in a capitalist system. This was the sort of opportunity that propelled MCI's successful assault on the Bell System's regulatory monopoly. It was the force that finally drove wireless beyond the control of the CPUC. Given these situations, entrepreneurs like William McGowan, Craig McCaw, and Sam Ginn searched relentlessly for weak spots in the regulatory regimes and often found ways to break the hold of the political system.

Both economic and political forces combined to push wireless away from governmental control and toward market control around the world. As the regulatory and state monopolies gave way, entrepreneurs rushed in to take advantage of the great opportunities and rewards available to successful innovators. As they did, the political regimes began to change. Regulators became market managers. Their old paradigm had been grounded in a desire to protect the consumer, and they did not abandon that goal. But they changed their means to that end. No longer trapped in an outdated historical context, they created a new paradigm that involved a primary emphasis on setting the rules for competition. Some of the critical ingredients of the rules were open access, transparency, nondiscrimination, and interconnection.

Governments did not wither away, of course. The politics of innovation continued to be of great importance in setting technical standards, and the European success in this regard provides an astonishing example of the rewards available for astute political innovation. In Europe, the government functioned outside the normal regulatory system and was able to be decisive about setting a workable standard. In the United States, the FCC refused to accept this role and left technological standards to the control of the market. Setting the standard in Europe fueled the process of change by providing a common platform for the kind of complementary innovations that normally follow after the introduction of a major new technology.[8] European standard setting indicated what can be accomplished when governments take a long-term, dynamic view of political economy.

Europe's success story makes the American regulatory and political performance look terrible. The United States in effect gave Europe a ten-year lead in wireless because of America's failure to set standards in a timely manner. In some respects, the ability of Vodafone, despite its smaller size, to acquire AirTouch can be attributed to the competitive advantages created by the European decision to standardize on GSM. Ginn and others argue that this view is short-sighted. The decision by the ITU (including Europe) to adopt an improved version of CDMA for third-generation technology reflects, Ginn says, the strength of the American model. "Superior technology doesn't always win in the marketplace," Ginn says, "but in this instance, technological competition produced superior technology."[9] Thus the industry debate continues, and all that is certain at this writing is that Europe won the first round in the race to develop a wireless world.[10]

The 3G auctions indicate almost as clearly what problems are created when governments take a short-term, static perspective on political economy. The auctions did exactly what the advisors had predicted: they maximized the returns to the government that was selling valuable spectrum; and they brought marginal prices and marginal costs close together, creating a situation for the competitors that approached the ideal of perfect competition. But that ideal was accompanied by some important unanticipated changes. The shift from beauty contests to auctions as a means of selling the radio spectrum generated income for every government that tried it, but it also ended the first, entrepreneurial phase of wireless expansion. With the auctions, the focus of change became consolidation. Costs were pushed up toward prices, and every business in the industry became attentive to the need to have a national footprint and the ability to generate economies of scale.[11] The 3G auctions left the companies with enormous debt, insufficient capital to build their systems, and ultimately the likelihood that prices for consumers will be higher. We see very little evidence that the goal of sustaining technological and operational innovation over the long term was served by this political innovation.

*

Although politics repeatedly trumped technology, in wireless there was clearly some co-evolution between governmental institutions and technical change. During the decades prior to World War II, technological and scientific constraints retarded innovation. Governments were under little pressure to change political regimes that favored hardwired systems that were either state-owned enterprises (SOEs) or pervasively regulated private companies such as AT&T. But following the war, new science and technology produced the opportunities that brought entrepreneurs flocking to wireless. Eventually they broke through and helped change the political landscape for the next generation of innovators in this industry and others, confirming the analyses of those scholars who have long maintained that new technologies normally force institutional reform on governments.[12] This was true in the United States as well as Europe, but the political changes were more substantial in Europe than they were in America. For the most part, Europe's telecom SOEs had lagged behind their U.S. counterparts, making the transition to wireless there more rapid and more complete. The technology in Europe favored markets over SOEs and fostered support for the European Union, which repeatedly demonstrated the advantages of a centralized approach to standardization in technical development.

In a more general sense, the new technologies of the Information Age and the economic growth they caused in the 1980s and 1990s put meat on the abstract bones of the conservative, market-oriented ideologies that were becoming dominant during those years. It was one thing to point out that the competitive model or the theory of comparative advantage would result in more economic progress for more people than would nationalistic controls or rate-of-return regulation. It was another thing, and a far more powerful argument, when you could point to a hand-held phone and the rapid spread of wireless as a specific product of the new age of competition and entrepreneurial freedom. The rapidity with which the industry moved from the first to the third (i.e., the present) generation of wireless technology also reinforced the market ideology. The process of technology transfer was now taking place faster than it ever had during the First and Second Industrial Revolutions.

*

Wireless was a global enterprise shaped by the risk-sharing proclivities of international lenders as well as national industrial policies. Capital constraints a la Schumpeter were foremost on the minds of entrepreneurs in every country. Sam Ginn's major decisions – for the spin, for the partnerships, and for the best means to achieve a national and international footprint – were all influenced in a significant way by the need to obtain financial support. Wireless devoured capital at a tremendous rate, and the fundraising was perforce global. No single nation could have possibly provided all of the funds needed to purchase the licenses, build the networks, operate the businesses, and then, suddenly, adjust to another new technology.

The complex global partnerships in wireless were a distinctive feature of business in the Third Industrial Revolution. No single firm could provide all of the capabilities needed to develop a successful wireless enterprise. The Second Industrial Revolution was dominated by large, vertically integrated organizations that wedded mass production and mass distribution. Firms that opted for the convergence strategy bet that the Information Age would see a similar integration of functions. But through the early entrepreneurial phase and even during the consolidation phase of wireless expansion, the pure-play and alliance strategies were the winners. Eventually, convergence may take place. But by early 2002, most of the firms that had adopted that strategy had either spun off their wireless operations or were looking for opportunities to do so.

Seen in their historical context, these developments mark a decisive shift in global political economy. The world's three industrial revolutions have been characterized by three phases of capitalism and three distinctive styles of innovation: individual entrepreneurship, hierarchical entrepreneurship, and now alliance entrepreneurship. As the story of Sam Ginn and AirTouch shows, wireless is a key example of the new era of alliance capitalism.[13] From the old Bell System, Pacific Telephone begat Pacific Telesis Group, which launched PacTel, which was transformed by the acquisition of Communications Industries and reshaped by the spin that created AirTouch. AirTouch became a global partner with operations around the world, all built to fit different markets, political systems, technologies, and financial settings. Then came the acquisition by Vodafone. All of these changes had taken place within sixteen years, and during these transitions, the boundaries of the firm had constantly shifted. Throughout, the opportunities and capabilities of the firm were governed, frequently contested, and then solidified through relationships with joint venture partners, government agencies, standard-setting bodies, suppliers, and agents.

Did creative destruction at this pace have a downside? Of course. It transformed the lives of many thousands of individuals, and some of these changes were intensely negative. This was the case with those employees who lost their jobs as a consequence of the consolidation movement in this industry. Even though Sam Ginn tried to protect his AirTouch employees, he could only soften the blows. He couldn't prevent them. Long before that happened, other workers in other firms as well as many of their managers and executives had been pushed around by the wireless revolution. This innovation, like all others, forced old forms of economic action to give way and left behind individuals and institutions unable or unwilling to take part in the new order.

Even those entrepreneurial types like Ginn and McCaw had some bitter along with the sweet in their lives. They were amply rewarded with the kind of capital gains that excite and frequently anger writers who are more concerned about economic equity than economic growth. So it has always been in successful capitalist societies. The bitter for both men came when the organizations to which they had devoted a substantial part of their lives disappeared in the consolidation era. The names were gone. New leaders devised new structures of authority and the existing organizational cultures quickly gave way. For most of us, that seems a small price to pay for the entrepreneurial returns that a CEO like Sam Ginn walked away with when AirTouch was acquired. But most of us are not entrepreneurs, not organizational builders, not the leaders who transform industries and societies. They leave their mark through the organizations they create, and they have every right to feel regret when those institutions disappear into another entrepreneur's creation.

Beyond regret in this case is the loss that the industry and the economy suffer when a unique, innovative, and efficient organization such as Air-Touch disappears. AirTouch was in many ways the archetypal business firm of this era. The company's powerful culture enabled it to maintain a dual hierarchy with tight financial controls and technological guidance (as in the Bell System) and a loose, multinational organization that encouraged local autonomy and innovation. Why, then, did AirTouch disappear into a global combine? – because assets were more important than management, organizational culture, or strategy in the consolidation phase of the industry, and because Vodafone, much favored by European investors, had the assets to buy out Ginn's creation instead of the other way around.

There is also a social downside to the wireless innovation. For some, wireless represents one of the most intrusive, socially unacceptable changes produced by the Information Age revolution. Whether it is loud conversations in trains, buses, and restaurants or wireless chats taking place on the street, the presence of the personal communicator has aroused substantial criticism. We hear the complaints almost every day: "I don't want to know about some stranger's business. I don't care about his or her arrangements for dinner tonight." What we are hearing is a very general critique about having our social space invaded by a stranger's wireless conversation. Responding to these complaints, the U.S. train service Amtrak has experimented with a "quiet car" in which wireless conversations are forbidden. Wireless has also given rise to health and safety concerns. It was rumored that wireless phones could cause cancer and that they increased the potential for accidents when used while driving. While there is no scientific evidence in support of the cancer hypothesis, there is abundant evidence indicating that drivers who are using their phones are indeed more likely to have accidents. Understandably, American states have begun to introduce legal restrictions on using a phone while driving. The ability to communicate anytime, anywhere has such broad appeal, however, that wireless has continued to rake in new customers despite these social issues.

*

Are there lessons to be learned from the AirTouch story and the entrepreneurial career of Sam Ginn and other Bell System executives who successfully made the transition to competition in wireless? Yes, we think there are, even though we are historians, not prophets. For business men and women, the wireless story underscores the dynamic character that economic life has had and will doubtless continue to have as the Third Industrial Revolution unfolds. Business institutions, no matter how well entrenched, can be quickly and radically transformed by changes in technology and by new regulatory, trade, and political regimes. The wireless story illustrates that the pace of change is accelerating; as it does, creative destruction on a global scale presents new problems for firms and for the millions of individuals who sustain their efforts.

When they solve these problems, "creative transformation" takes place, and wireless abounds with examples of organizational adaptation to new market, technological, and political settings. Creative transformation accounts to a considerable degree for the continued importance of the Bell culture and Bell organizations to the industry. Although these businesses jettisoned many individuals, well-established values, and traditional strategies, they were successful when they were able – in the AirTouch style – to blend the Bell sources of excellence with a market-grounded sensitivity to costs, customers, and competitors. Any business facing either deregulation or radical technological change would do well to examine the successful examples of organizational transformation in wireless.

Successful managers and executives have always been sensitive to their economic and political environments, but the era of alliance capitalism imposes new demands for flexibility. The boundaries of the firm are becoming blurred as suppliers, customers, and outside institutions are more closely integrated into the day-to-day operations of the business. This new era requires greater openness to outside ideas and increased flexibility in dealing with different managerial perspectives and cultural expectations. AirTouch's management – perhaps because of its roots in the multicultural society of California – demonstrated a remarkable ability to assimilate outside ideas, from the merger with Communications Industries to the forging of its new businesses in cultures as different as those in Germany and Japan.

The AirTouch story also suggests that Information Age business leaders will find it essential to increase the amount of thought they give to managing their company's most valuable resource – its people. Although the twentieth-century evolution from industrial management to human resources has already involved a shift in this direction, the new era epitomized by wireless seems to call for more radical change. In the evolution from AT&T's AMPS service to the creation and development of AirTouch, Sam Ginn and his team transformed the character of the company's relationship with its employees. Ginn, Sarin, Lee Cox, and others cultivated a partnership with employees. They eschewed the adversarial history of organized labor and sought instead to motivate employees by giving them a stake in the business through the two-by-four program. Certainly, their program was tame by comparison with the form of stock-option compensation that became popular at the height of the Internet bubble in the spring of 2000.[14] But the AirTouch plan was a radical departure from the traditions of the Bell System. At the same time, Ginn and his colleagues reaffirmed the importance of employees as people by borrowing from the traditions of the Bell System to focus on their professional development. They thus engendered tremendous loyalty at all levels of the organization and created an environment that encouraged innovation throughout the business.

For scholars interested in business strategy and organizational theory, the AirTouch story offers some important lessons and intriguing questions. Throughout the development of wireless, the innovators all underestimated the potential of the market and overestimated their ability to acquire the capital they needed to realize their entrepreneurial objectives. Long accustomed to intellectuals' sarcastic references to business boosterism, we are surprised by the failure to appreciate fully the desire for anytime–anywhere communication. Perhaps what we need is a more refined concept of the early, high-risk stage of innovation. Then, there seems to be a premium on underestimating demand in a manner that is probably influenced by the investor community's desire to reduce risk by staying in touch with what is familiar. When you are trying to raise capital, as all of the wireless innovators were, you don't want to be laughed out of the room. While the entrepreneurs' estimates of future demand were thus too conservative, they were extremely radical compared to those of the established players in telecommunications. During the early phase of wireless development, MCI made several abortive attempts at entry, but AT&T stayed on the sidelines. Even after buying McCaw's interests, AT&T was slow to go international. In this case, the firm's commitment to a strategy of wireline consolidation apparently prevented it from appreciating either the potential demand or the advantages it had when licenses were being awarded through "beauty contests."

Of course, even AT&T was eventually unable to muster all of the capital it needed to complete the implementation of its consolidation strategy. In that regard, it followed into trouble most of the businesses that started or attempted to start wireless enterprises. The most outstanding example was McCaw Cellular Communications, which AT&T bought as it tried to come from behind in wireless. AirTouch hit the wall on capital when it tried to mount expensive ventures in Germany and Britain at the same time. These various encounters with capital constraints indicate that there is a fundamental conflict between an entrepreneurial strategy and the strategic need for a trigger point that will slow or even end the capital-burning phase of growth. Since the entrepreneurial temperament and vision are inherently optimistic and opposed to this sort of limit, the problem for the strategist involves the personal aspects of leadership and the organizational requirements for an effective trigger, an institutional cutoff on capital-burning expansion. Central to their planning should be an appreciation of the power wielded by the investment community and the new role played by the global market for corporate control. Both enhance the value of having a workable trigger.

The history of wireless points as well to the need for strategists and organizational theorists to give more consideration to the role of political institutions in shaping market environments. The ability to gain competitive advantage from a superior understanding of the political dynamics affecting an industry can often be critical to success – especially in an industry (like telecommunications) experiencing a historic shift from state ownership or strong regulation to market competition. In other leading Third Industrial Revolution industries, too, political strategies can be as decisive as new technology in shaping competitive outcomes. AirTouch's strategy for global development was guided as much by political as by economic or technical considerations. Vodafone's march to international leadership was based on a similar strategy.

The wireless experience also provides some lessons that nation states and the emerging economic regions in Europe, the Western Hemisphere, and Asia could well take to heart. To planners and politicians in the United States, it would be good to consider how the nation acquired the best telecommunications system in the world and then quickly lost its advantage in wireless. The speed with which this change took place should cause some concern, as should the political process that European nations used to create competitive advantage. Americans – proud of their history and economic power – have always learned from other nations rather grudgingly. This is understandable in a nation that had, by the end of the nineteenth century, already developed the leading industrial economy in the world. But in the fast-moving Information Age setting, it might be wise to swallow some national pride and attempt in future years to create national and state political capabilities equal to the business capabilities that U.S. corporations have demonstrated in the past two decades of creative transformation.

That, it seems, is what Britain has done – much to its advantage. Vodafone's success mirrors a U.K. business recovery that is one of the great stories of the Third Industrial Revolution. Britain fell far behind Germany and the United States during the Second Industrial Revolution, when British companies failed to keep up with new electrochemical technologies and methods of mass production. But the British business system continued to have great strength in services. In wireless, Vodafone has become the industry's leader by dint of an outstanding performance in both the entrepreneurial and consolidation eras. The acquisition of Mannesmann was an appropriate symbol of British success.

The struggle over Mannesmann provided an excellent guide to the problems Germany has experienced in recent years in high-tech industries like telecommunications. Where a deep-set culture is involved, as it is in Germany, change can be especially painful and politically volatile. Perfectly happy to take advantage of Britain's wide-open market for corporate control, Mannesmann turned around and mounted an unsuccessful (even embarrassing) resistance to Vodafone's impressive buyout offer. Instead of becoming full-fledged global players, the German executives wanted to dabble in world markets without embracing them. That is not a sustainable position in the wireless world. Japan will probably have to learn the same lesson in the same way at some time in the near future. In this industry, economic nationalism began to give way when governments started to abandon their state-owned and pervasively regulated telecoms. To date, the drift is still toward competition in relatively open, global markets. The resulting global oligopolies will be hard for many citizens in democracies to accept, but they are likely to find that outcome more tolerable than the economic price of resistance to innovation. With digital technology and a link between the wireless phone and the Internet, this service industry will punch another hole in the national boundaries that dominated economic growth during the Second Industrial Revolution.

By bringing our readers up close to these events – through the history of wireless, the activities of AirTouch, and the entrepreneurship of Sam Ginn and others – we have tried to open a window on innovation, the driving force of capitalism. Wireless unleashed successive waves of creative destruction and creative transformation in the global capitalist economy. AirTouch and Sam Ginn played an important role in that process, and they earned their place in the history of wireless and of the Third Industrial Revolution of our time. They demonstrated once again why the ability of the capitalist system to respond to a changing environment, to absorb new technologies, and to generate new ways of doing business is what distinguishes it from alternative forms of political economy. Entrepreneurship in wireless created new organizations, new jobs, wealth, income, and, above all, new services that people around the world wanted. Today, millions are able to access a world of information from a device that fits in the palm of their hand. This new technology satisfies two age-old human desires: to retrieve the information we need in the moment, and to communicate with those people who are foremost in our minds anytime, anywhere around the world.

Notes

CHAPTER ONE

1. Interview with Sam Ginn, March 23, 2000.
2. Kenny Moore, "They Got Off on the Right Track," *Sports Illustrated*, August 13, 1984, p. 60; Peter Ueberroth with Richard Levin and Amy Quinn, *Made in America* (New York, 1985), p. 19; Gary Smith, "I Do What I Want to Do," *Sports Illustrated*, July 18, 1984, p. 22.
3. Joseph A. Schumpeter, *The Theory of Economic Development* (Cambridge, MA, 1934), pp. 84–90.
4. Joseph A. Schumpeter, *Capitalism, Socialism and Democracy* (New York, 1942). Schumpeter died in 1950 in Connecticut.
5. See, for example, Richard Nelson and Sidney Winter, *An Evolutionary Theory of Economic Change* (Cambridge, 1982), which the authors characterize as a "neo-Schumpeterian" approach; Lars Magnusson (Ed.), *Evolutionary and Neo-Schumpeterian Approaches to Economics* (Boston, 1994); Richard R. Nelson, "Capitalism as an Engine of Progress," *Research Policy* 19 (1990), pp. 193–214; and Giovanni Dosi et al., *Technical Change and Economic Theory* (London, 1988). For a more popular example, see S. Steinberg, "Schumpeter's Lesson: What Really Happened in Digital Technology in the Past Five Years," *Wired*, January 1998.
6. Alfred D. Chandler, Jr., *Scale and Scope: The Dynamics of Modern Capitalism* (Cambridge, MA, 1990); Richard R. Nelson (Ed.), *National Innovation Systems* (New York, 1993); Oliver E. Williamson, *The Economic Institutions of Capitalism: Firms, Markets, Relational Contracting* (New York, 1985); Giovanni Dosi et al. (Eds.), *Technology, Organization, and Competitiveness* (New York, 1998); David Hounshell, "The Evolution of Industrial Research in the United States," in Richard S. Rosenbloom and William J. Spencer (Eds.), *Engines of Innovation: U.S. Industrial Research at the End of an Era* (Boston, 1996), pp. 13–87.
7. O. Casey Corr, *Money from Thin Air: The Story of Craig McCaw, the Visionary Who Invented the Cell Phone Industry, and His Next Billion-Dollar Idea* (New York, 2000); James B. Murray, Jr., *Wireless Nation: The Frenzied Launch of the Cellular Revolution in America* (Cambridge, MA, 2001).
8. In this context, we also offer some perspective on the dynamic character and potential of large-scale bureaucratic organizations and on the transformation of industrial

corporations in the transition from the Second to the Third Industrial Revolution. Prior to the 1984 breakup, AT&T was a variation on the classic Chandlerian multidivisional or M-form organization. After the breakup, Ma Bell and her offspring suddenly had to find structures and strategies appropriate to the technological and political changes associated with the Information Age. On bureaucracy, see Max Weber, *The Theory of Social and Economic Organization* (New York, 1947); Anthony Downs, *Inside Bureaucracy* (Boston, 1966); Martin Albro, *Bureaucracy* (New York, 1970); and Michel Crozier, *The Bureaucratic Phenomenon* (Chicago, 1964). On the Chandlerian organization, see Alfred D. Chandler, Jr., *Strategy and Structure: Chapters in the History of the Industrial Enterprise* (Cambridge, MA, 1962); *The Visible Hand: The Managerial Revolution in American Business* (Cambridge, MA, 1977); and *Scale and Scope.*

9. Ernest Braun and Stuart Macdonald, *Revolution in Miniature: The History and Impact of Semiconductor Electronics,* 2nd ed. (New York, 1982). See also Nicholas Negroponte, *Being Digital* (New York, 1995), and George F. Gilder, *Microcosm: The Quantum Revolution in Economics and Technology* (New York, 1989).

10. Pier Angelo Toninelli, *The Rise and Fall of State-Owned Enterprise in the Western World* (New York, 2000).

11. David Held et al., *Global Transformations: Politics, Economics and Culture* (Stanford, 1999); Thomas L. Friedman, *The Lexus and the Olive Tree: Understanding Globalization* (New York, 1999).

12. See for example Martin Kenney, *Understanding Silicon Valley: The Anatomy of an Entrepreneurial Region* (Stanford, 2000); AnnaLee Saxenian, *Regional Advantage: Culture and Competition in Silicon Valley and Route 128* (Cambridge, MA, 1994); Dan Steinbock, "Assessing Finland's Wireless Valley: Can the Pioneering Continue?" *Telecommunications Policy* 25 (2001), pp. 71–100; and Michael Porter, *The Competitive Advantage of Nations* (New York, 1990), pp. 65–7.

CHAPTER TWO

1. The Bell System was similar in most regards to the other large, multidivisional firms typical of the Second Industrial Revolution; as Alfred Chandler has observed, that revolution was led by professional managers rather than owners. See Alfred D. Chandler, Jr., *The Visible Hand: The Managerial Revolution in American Business* (Cambridge, MA, 1977). For the history of the early years of the Bell System, see George David Smith, *The Anatomy of a Business Strategy: Bell, Western Electric, and the Origins of the American Telephone Industry* (Baltimore, 1985), and Robert Garnet, *The Telephone Enterprise: The Evolution of the Bell System's Horizontal Structure, 1876–1909* (Baltimore, 1985). The most powerful critique of the Bell System and its authority prior to 1960 was offered by N. R. Danielian, *A.T.&T.: The Story of Industrial Conquest* (New York, 1939).

2. The father, who was an AT&T recruiter, met Sam Ginn when he visited their home in Atlanta. The father was sufficiently impressed that weekend to pitch the Bell System to Sam, who was attracted by the possibility of working for the largest private corporation in the world.

3. As early as the 1920s, AT&T had begun to focus management recruiting on college graduates and to study correlations between academic and extracurricular achievement in college and management performance. By the 1950s, the company had begun to formalize management development programs, relying on social science

research. See Douglas W. Bray et al., *Formative Years in Business: A Long-Term AT&T Study of Managerial Lives* (New York, 1974).

4. In 1956, AT&T had initiated a management progress study that developed a "high-risk, high-reward" fast-track management program for college graduates. For a longitudinal study of 813 AT&T managers, see Ann Howard and Douglas W. Bray, *Managerial Lives in Transition: Advancing Age and Changing Times* (New York, 1988).

5. Herzberg taught psychology at Western Reserve University in the late 1950s and focused his research on job attitudes and motivation. See Frederick Herzberg et al., *The Motivation to Work* (New York, 1959) and *Job Attitudes: Review of Research and Opinion* (Pittsburgh, 1957). Interview with Dwight Jasmann, May 2, 2000.

6. For a discussion of the role of women in the early Bell System, see Kenneth Lipartito, "When Women Were Switches: Technology, Work and Gender in the Telephone Industry, 1890–1920," *American Historical Review* 99 (October 1994), pp. 1075–1111.

7. For the educational and family backgrounds of management trainees, see Howard and Bray, *Managerial Lives in Transition,* pp. 27–33. Interview with Sam Ginn, February 17, 2000.

8. Peter Temin with Louis Galambos, *The Fall of the Bell System* (Cambridge, 1987), pp. 160–75.

9. On the values of the Bell System, see Alvin Von Auw, *Heritage & Destiny* (New York, 1983), and W. Brooke Tunstall, *Disconnecting Parties: Managing the Bell System Break-Up* (New York, 1985).

10. Smith, *The Anatomy of a Business Strategy*; Garnet, *The Telephone Enterprise.*

11. Stephen B. Adams and Orville R. Butler, *Manufacturing the Future: A History of Western Electric* (Cambridge, 1999).

12. Neil H. Wasserman, *From Invention to Innovation: Long-Distance Telephone Transmission at the Turn of the Century* (Baltimore, 1985). See also John Brooks, *Telephone: The First Hundred Years* (New York, 1976).

13. Leonard Reich, *The Making of American Industrial Research: Science and Business at GE and Bell, 1876–1926* (Cambridge, 1985).

14. Louis Galambos, "Theodore N. Vail and the Role of Innovation in the Modern Bell System," *Business History Review* 66 (Spring 1992), pp. 95–126.

15. Robert Bornholz and David S. Evans, "The Early History of Competition in the Telephone Industry," in David S. Evans (Ed.), *Breaking Up Bell: Essays on Industrial Organization and Regulation* (New York, 1983); Richard Gabel, "The Early Competitive Era in Telephone Communications, 1893–1920," *Law and Contemporary Problems* 34 (Spring 1969), pp. 340–69.

16. Garnet, *The Telephone Enterprise.*

17. Nathan Kingsbury handled the negotiations with the U.S. Justice Department on behalf of AT&T. AT&T's commitment to universal service was embedded in a letter to the Attorney General dated December 16, 1913: *Letter to the Attorney General* (Washington, DC: Government Printing Office, 1914). Cited in Alan Stone, *Public Service Liberalism: Telecommunications and Transitions in Public Policy* (Princeton, 1991), pp. 191–5.

18. On the evolution of telephone systems in Europe, see Eli Noam, *Telecommunications in Europe* (New York, 1992).

19. *United States v. Western Electric Co.,* Trade Cas. (Consent Decree) (CCH) 68,246 (D.N.J. 1956). As author Kevin G. Wilson points out, the consent decree was ironically labeled the "Final Judgment." The breakup of the Bell System in 1984 took

place under an agreement to modify this Final Judgment. See Kevin G. Wilson, *Deregulating Telecommunications: U.S. and Canadian Telecommunications, 1840–1997* (Lanham, MD, 2000).

20. FCC Docket 8658, cited in Dale N. Hatfield, "FCC Regulation of Land Mobile Radio – A Case History," in Leonard Lewin (Ed.), *Telecommunications: An Interdisciplinary Text* (Dedham, MA, 1984), p. 108.

21. AT&T completed the first transcontinental microwave network, using 107 relay stations spaced roughly thirty miles apart, to form a communications link from New York to San Francisco in 1951. Within three years, the company had over 400 microwave stations operating around the country and, by 1958, microwave accounted for nearly a quarter of the nation's long distance circuits. Gerald Brock, *The Telecommunications Industry* (Cambridge, MA, 1981), pp. 181–7.

22. Federal Communications Commission, *In the Matter of Allocation of Microwave Frequencies in the Band Above 890 Mc.,* 27 FCC 359 (1959).

23. Ralph Nader and William Taylor, *The Big Boys: Power and Position in American Business* (New York, 1986), p. 425. On the history of MCI, see Philip L. Cantelon, *The History of MCI: The Early Years* (Dallas, 1993), and Larry Kahaner, *On the Line* (New York, 1986). For the FCC's decisions, see Federal Communications Commission, *In Re Applications of Microwave Communications, Inc.,* 18 FCC 2d 953 (1969), recon. denied, 21 FCC 2d 825 (1970), at 965.

24. Temin with Galambos, *The Fall of the Bell System,* p. 42. Refer to *Carter v. AT&T,* F. Supp. 188, 192 (N.D. Tex. 1966), aff'd., 365 F. 2d 486 (5th Cir. 1966), cert. denied, 385 U.S. 1008, 1967.

25. AT&T was not blind to the changes taking place. In 1968, aware that the telecommunications industry was about "to be revolutionized by new technologies, new social attitudes, [and] new government attitudes," the company hired Alvin Toffler to think about the role of the company in the future. Few, if any, changes resulted from Toffler's report, however. Alvin Toffler, *The Adaptive Corporation* (New York, 1985).

26. Cantelon, *The History of MCI.*

27. Temin with Galambos, *The Fall of the Bell System,* p. 50; Alan Stone, *Wrong Number: The Breakup of AT&T* (New York, 1989), p. 181; Steve Coll, *The Deal of the Century: The Breakup of AT&T* (New York, 1986), p. 15.

28. Competition in long distance forced the FCC to adjust prices closer to costs in the 1970s. After MCI won a legal battle allowing it to provide the equivalent of message toll service for long-distance calls, the ENFIA tariff was adopted in 1978 to set charges for access to AT&T's local exchanges for competitive carriers like MCI. After divestiture, the FCC used access charges to replace the traditional subsidy that long distance had provided to local services through the separations process. See Richard H. K. Vietor, "AT&T and the Public Good: Regulation and Competition in Telecommunications, 1910–1987," in Stephen P. Bradley and Jerry A. Hausman, *Future Competition in Telecommunications* (Boston, 1989), pp. 85–9. Also, Temin with Galambos, *The Fall of the Bell System,* pp. 131–2. Interview with Sam Ginn, February 17, 2000; interview with Sam Ginn, August 14, 1991.

29. Stone, *Wrong Number;* Fred W. Henck and Bernard Strassburg, *A Slippery Slope: The Long Road to the Breakup of AT&T* (New York, 1988); Gerald R. Faulhaber, *Telecommunications in Turmoil: Technology and Public Policy* (Cambridge, MA, 1987).

30. Temin with Galambos, *The Fall of the Bell System,* pp. 220–1.

CHAPTER THREE

1. Interview with Philip J. Quigley, May 7, 1992. For background on the role of marketing in the Bell System and efforts to invigorate AT&T's efforts in the 1960s and 1970s, see Alvin Von Auw, *Heritage & Destiny* (New York, 1983), pp. 162–3, and Peter Temin with Louis Galambos, *The Fall of the Bell System* (Cambridge, 1987).

2. Hugh G. J. Aitken, *Syntony and Spark – The Origins of Radio* (Princeton, 1976), chap. 5.

3. W. Rupert MacLaurin, *Invention & Innovation in the Radio Industry* (New York, 1949), pp. 35–50.

4. Hugh Aitken, *The Continuous Wave: Technology and American Radio, 1900–1932* (Princeton, 1985).

5. Daniel Levinthal has pointed out that Fessenden's failure in wireless telephony and broadcast radio resulted from the lack of a necessary technological innovation (a continuous wave transmitter) and "equally important, an inability to identify and pursue a market that would support [his] efforts." Daniel A. Levinthal, "The Slow Pace of Rapid Technological Change: Gradualism and Punctuation in Technological Change," *Industrial and Corporate Change* 7, no. 2 (1998), pp. 217–47.

6. MacLaurin, *Invention & Innovation in the Radio Industry*, pp. 79–82.

7. Stephen B. Adams and Orville R. Butler, *Manufacturing the Future: A History of Western Electric* (Cambridge, 1999), p. 4; MacLaurin, *Invention & Innovation in the Radio Industry*, pp. 84–5.

8. MacLaurin, ibid., pp. 79–82.

9. Daniel E. Noble details the key role police departments played in developing land-mobile radio communications in "The History of Land-Mobile Radio Communications," *Proceedings of the IRE* 50, no. 5 (1962), pp. 1405–14. See also Heidi Kargman, "Land Mobile Communications: The Historical Roots," in Raymond Bowers et al. (Eds.), *Communications for a Mobile Society: An Assessment of New Technology* (Beverly Hills, 1978), p. 25.

10. These systems relied on amplitude modulation (AM) transmission. But with a mobile receiver, static and "flutter" created transmission problems. The development of FM systems in the late 1930s improved the range and quality of communications considerably. See Noble, "The History of Land-Mobile Radio Communications," p. 1407.

11. Committee on Evolution of Untethered Communications, *The Evolution of Untethered Communications* (Washington, DC, 1997), p. 18.

12. Garry A. Garrard, *Cellular Communications: Worldwide Market Development* (Boston, 1998), p. 14.

13. Kathi Ann Brown, *Critical Connection: The Motorola Service Station Story* (Rolling Meadows, 1992), pp. 64–5.

14. Robert Horvitz, "Personal Radio," *Whole Earth Review* (Spring 1986), p. 34.

15. This new service was known in the Bell System as Improved Mobile Telephone Service (IMTS). Garrard, *Cellular Communications*, p. 19.

16. Duane L. Huff, "Cellular Radio," *Technology Review*, Nov.–Dec. 1983, p. 53.

17. General Mobile Radio Service, 13 FCC (1949), pp. 1193–1212, cited in John W. Berresford, "The Impact of Law and Regulation on Technology: The Case History of Cellular Radio," *Business Lawyer*, May 1989.

18. Clayton Niles, unpublished autobiography, draft dated August 23, 2000.

19. David Scott, "Journey Through the Past," *Telocator*, November 1988, pp. 48–53.

20. Michael Paetsch, *Mobile Communications in the U.S. and Europe: Regulation, Technology and Markets* (Boston, 1993), p. 120.
21. Eric Schuster to the authors, December 4, 2001. See also Brown, *Critical Connection*.
22. S. Warner, "Truly Portable Mobile Telephones One of Latest Steps in Long Evolution of Phone," *Communication News,* August 1983, pp. 48–9. See also Joseph G. Morone, *Winning in High-Tech Markets: The Role of General Management* (Boston, 1993), p. 73.
23. Morone, ibid., p. 74.
24. CB applications took off early in 1975. The peak of the craze came in late 1976, when applications spiked upward and reached a million in a single month. See Cary Hershey et al., "Personal Uses of Mobile Communications: Citizens Band Radio and the Local Community," in Bowers, *Communications for a Mobile Society,* p. 233.
25. D. Wiley, "World Communications Year ... 1983 ... Will Be Landmark Year in USA," *Communications News,* January 1983, pp. 26–41; "M-C Expo 83 Examines Mobile Communications Industry," *Communications News,* April 1983, p. 94; Robert T. Stovall, "Cellular Talk," *Financial World,* January 31, 1983, p. 70; Lehman Brothers Kuhn Loeb Research, "Cellular Mobile Radio Telecommunications Service: Applications for the Top 30 Markets," September 24, 1982.
26. Harry E. Young, "A Page in History: One Man's Ideas 50 Years Ago Helped Create a Multi-Billion Dollar Wireless Communications Industry," *Communications,* September 1995, pp. 51–4, and October 1995, pp. 31–3.
27. Morone, *Winning in High-Tech Markets,* pp. 65–124.
28. Michael K. Kellogg et al., *Federal Telecommunications Law* (Boston, 1992), p. 642.
29. Horvitz, "Personal Radio," p. 34.
30. FCC Docket 8976, referenced in Mark V. Nadel et al., "Land Mobile Communications and the Regulatory Process," in Bowers, *Communications for a Mobile Society,* p. 69.
31. United States General Accounting Office, "Concerns about Competition in the Cellular Telephone Service Industry," July 1992.
32. Raj Pandya, *Mobile and Personal Communication Services and Systems* (New York, 2000), p. 15.
33. General Accounting Office, "Concerns about Competition in the Cellular Telephone Service Industry," p. 14.
34. Eric Schuster to the authors, December 4, 2001.
35. "A Mobile Future," *The Economist,* October 13–19, 2001, p. 19.
36. The FCC required AT&T to offer cellular service through a separate wholly owned subsidiary, which AT&T established as AMPS on May 5, 1980. AT&T later merged AMPS into a new company, AT&T Cellular Company. See 11 CPUC 2d at 838.
37. Alexander L. Taylor, "Why So Many Phones Are Going 'Beep ...'," *Time,* April 11, 1983, p. 75; Huff, "Cellular Radio," p. 53.
38. Jules Millman, President, Alladin's Castle, Inc., to Robert Ford, Illinois Bell, April 6, 1977, in AT&T, "High Capacity Mobile Telecommunications System: Developmental System Report," June 8, 1977.
39. Miriam L. Schmitt, Harding Real Estate, to Joseph Enenbach, Illinois Bell, March 14, 1977, in AT&T, "High Capacity Mobile Telecommunications System."
40. Interview with Thomas H. Ehler, August 8, 2000.
41. Garrard, *Cellular Communications,* p. 31.
42. Interview with Thomas H. Ehler, August 8, 2000.

43. AT&T, "High Capacity Mobile Telecommunications System: Developmental System Report," June 8, 1977.

CHAPTER FOUR

1. Interview with Craig McCaw, June 8, 2000.
2. David Scott, "Journey Through the Past," *Telocator,* November 1988, pp. 48–53.
3. *Amendment of Part 21 of the Commission's Rules,* 12 FCC 2d, p. 846.
4. Federal Communications Commission, *Land Mobile Service Notice, Use of 806–960 MHz Band,* 14 FCC 2d, pp. 311–17.
5. Broadcasters and the Bell System had been fighting over spectrum since the late 1940s. Repeatedly, the Bell System's appeals for spectrum for land mobile radio had been denied by the FCC in favor of television. See Dale N. Hatfield, "FCC Regulation of Land Mobile Radio – A Case History," in Leonard Lewin (Ed.), *Telecommunications: An Interdisciplinary Text* (Dedham, MA, 1984), pp. 109–10.
6. Henry G. Fischer and John W. Willis (Eds.), *Pike and Fischer Radio Regulations,* 2nd series, vol. 19, pp. 1590–2. See also Mark V. Nadel et al., "Land Mobile Communications and the Regulatory Process," in Raymond Bowers et al. (Eds.), *Communications for a Mobile Society: An Assessment of New Technology* (Beverly Hills, 1978), p. 69; Dale N. Hatfield, "FCC Regulation of Land Mobile Radio – A Case History," in Lewin, *Telecommunications,* pp. 105–33; and John W. Berresford, "The Impact of Law and Regulation on Technology: The Case History of Cellular Radio," *Business Lawyer,* May 1989.
7. Fischer and Willis, *Pike and Fischer Radio Regulations,* 2nd series, vol. 33, pp. 466–8; Nadel, "Land Mobile Communications and the Regulatory Process," p. 72.
8. Joseph G. Morone, *Winning in High-Tech Markets: The Role of General Management* (Boston, 1993), pp. 75–7.
9. John Mitchell, as quoted in Morone, *Winning in High-Tech Markets,* p. 75.
10. *Second Memorandum Opinion and Order,* Docket 18262, July 28, 1971 (31 FCC 2d 50).
11. Clayton Niles, unpublished autobiography, draft dated August 23, 2000.
12. Nadel, "Land Mobile Communications and the Regulatory Process."
13. Clay T. Whitehead to Hon. Dean Burch, August 17, 1973, Memorandum, Office of Telecommunications, Executive Office of the President, Washington, DC, p. 2, referenced in Nadel, "Land Mobile Communications and the Regulatory Process," p. 77.
14. Fischer and Willis, *Pike and Fischer Radio Regulations,* 2nd series, vol. 33, pp. 466–8.
15. *Second Report and Order,* Docket 18262, May 1, 1974 (46 FCC 2d 75).
16. 525 F. 2d 630 (D.C. Circ., 1976), referenced in Nadel, "Land Mobile Communications and the Regulatory Process," pp. 72–3.
17. Memorandum Opinion and Order, Docket 18262, March 19, 1975 (51 FCC 2d 945).
18. Duane L. Huff, "Cellular Radio," *Technology Review,* Nov.–Dec. 1983, p. 53; Lehman Brothers Kuhn Loeb Research, "Cellular Mobile Radio Telecommunications Service: Applications for the Top 30 Markets," September 24, 1982; GAO, "Concerns about Competition in the Cellular Telephone Service Industry," July 1992, p. 14.
19. Jeffrey Frey and Alfred M. Lee, "Technologies for Land Mobile Communications 900-MHz Systems," in Bowers, *Communications for a Mobile Society,* p. 59.

20. Morone, *Winning in High-Tech Markets,* pp. 75–7.
21. Greg More, "Regulators Mold Cellular Market," *Telephone Engineer & Management,* August 1, 1984, p. 126.
22. O. Casey Corr, *Money from Thin Air* (New York, 2000), p. 93.
23. "Cellular Mobile Radio Service Bringing Many New Opportunities for Small Telcos," *Communication News,* May 1983, p. 134; Lehman Brothers Kuhn Loeb Research, "Cellular Mobile Radio Telecommunications Service."
24. The agreement, signed on October 26, 1982, included AMPS, GTE Mobilnet, Inc., Continental Mobilcom, Inc., and United States Cellular Corporation. The agreement was approved by the FCC on March 31, 1983. See 11 CPUC 2d at 838.
25. Corr, *Money from Thin Air,* p. 104.
26. "Mobile Telephones: Beginning of a Boom," *Nation's Business,* March 1983, p. 68; " 'Third Tier' Cellular Filings Total 576," *Telephone Engineer & Management,* April 15, 1983, p. 41; Lehman Brothers Kuhn Loeb Research, "Cellular Mobile Radio Telecommunications Service. "
27. Clayton Niles to the authors, December 14, 2001.
28. Fleming Meeks, "Cellular Suckers," *Forbes,* July 25, 1988, p. 41.
29. Michael Kinsley, "Hello, Hillary," *The New Republic,* May 2, 1994, p. 6.
30. James B. Murray, Jr., *Wireless Nation: The Frenzied Launch of the Cellular Revolution in America* (Cambridge, MA, 2001).

CHAPTER FIVE

1. Interview with Sam Ginn, August 14, 1991. Interview with Sy Graff, December 12, 1991.
2. Interview with Sam Ginn, August 14, 1991. Interview with Sam Ginn by Tracey Panek, September 27, 1999.
3. American Telephone and Telegraph Company, *Telephone Statistics of the World* (New York, 1912), p. 20.
4. Eric John Abrahamson, "Keeping the Wolves at Bay: Regulation in the Telephone Industry in Los Angeles and California, 1902–1917," paper presented at Johns Hopkins University, April 1995. Alan Stone, *Public Service Liberalism: Telecommunications and Transitions in Public Policy* (Princeton, 1991), pp. 191–5.
5. Board of Public Utilities, City of Los Angeles, *Annual Report,* 1916–1917.
6. The California Railroad Commission (CRC) had received authority to modify rates and freight classification schedules without waiting for complaints from shippers. It could also set absolute rates, rather than just maximum rates. At the same time, the CRC was given the authority to regulate all utilities in California, including telephone companies. California Railroad Commission, *Report,* 1911–1912, p. 91. See also Mansel Blackford, *The Politics of Business in California, 1890–1920* (Columbus, 1977), chap. 5.
7. California Railroad Commission, *Decision 1655, Town of Antioch v. P.G. & E.,* CRC 9 (July 6, 1914). California Public Utilities Commission, *Decision 41416, Pacific Telephone & Telegraph Co.,* 48 CPUC 1 (April 6, 1948).
8. California Railroad Commission, *Decision 28764, City of Los Angeles v. Southern California Tel. Co.,* 39 CRC 739 (April 27, 1936).
9. Richard Gabel, *Development of Separations Principles in the Telephone Industry* (East Lansing, MI, 1967); James W. Sichter, "Separations Procedures in the Telephone Industry: The Historical Origins of a Public Policy," Harvard University

Program on Information Resources Policy, Publication P-77-2, Cambridge, MA, January 1977.

10. Interview with James McCraney and Paul Popenoe, February 23, 1992.

11. Ibid.

12. California Public Utilities Commission, *Decision 74917, In Re The Pacific Telephone & Telegraph Co.,* 69 CPUC 53 (November 6, 1968).

13. Section 441, *Tax Reform Act of 1969.*

14. California Public Utilities Commission, *Decision 78851, In Re Pacific Telephone & Telegraph Co.,* 72 CPUC 327 (June 22, 1971). California Supreme Court, *City of Los Angeles v. P.U.C., 7.* Cal 3d 331 (1972).

15. Eric John Abrahamson with Marjorie Wilkens, *Learning to Compete: A History of Pacific Telesis Group* (San Francisco, 1994). SBC Communications Archives.

16. On the New York service crisis, see Peter Temin with Louis Galambos, *The Fall of the Bell System* (Cambridge, 1987), pp. 55–7.

17. California Public Utilities Commission, *Decision 90642, In Re Pacific Telephone & Telegraph Co.,* 2 CPUC 2d 89 (1979).

18. Interview with Robert Dalenberg, January 15, 1992; interview with John Bryson, April 17, 1992.

19. California Public Utilities Commission, *Decision 91495, In Re Pacific Telephone & Telegraph Co.,* 3 CPUC 2d 401 (1980).

20. California Public Utilities Commission, *Decision No. 93367,* 6 CPUC 2d (August 4, 1981).

21. "Argument of Mr. Guinn," Before the Public Utilities Commission of the State of California, *Application No. 59849,* June 30, 1981, vol. 86, pp. 9458–9734.

22. Robert Bartlett, "Phone Company's Rate Request Challenged," *San Francisco Chronicle,* July 1, 1981, p. 6.

23. Some people asserted that this rate of return was extravagant, but with interest rates hovering near 20 percent, the commission said it was reasonable. California Public Utilities Commission, *Decision No. 93367, In Re Pacific Telephone & Telegraph Co.,* 6 CPUC 2d (August 4, 1981). See also *San Francisco Chronicle,* August 5, 1981, p. 1.

24. Harre W. Demoro, "How AT&T Plans to Shed Pac Tel," *San Francisco Chronicle,* February 3, 1982.

25. AT&T, "Reply Comments to the American Telephone and Telegraph Company," *U.S. v. AT&T,* Civil Action No. 82-0192, U.S. District Court for the District of Columbia, May 21, 1982, pp. 76–7.

26. U.S. District Court for the District of Columbia, *U.S. v. AT&T,* Civil Action No. 82-0192, "Modification of Final Judgment."

27. Guinn charged that AT&T's Plan of Reorganization left Pacific in a weak financial position: its cost of debt would be far higher than the Bell System average – 9.9 percent compared with 8.6 percent. AT&T's decision to treat Pacific's $553 million in nonvoting, fixed-charge preferred stock as equity would leave the company with an effective debt ratio of 54.4 percent, well above the 45-percent level of the other Baby Bells. Moreover, the continuing burden of the depreciation-related tax liability threatened to exacerbate Pacific's financial situation, and AT&T seemed poised to walk away from that issue. "Affidavit of Donald Guinn," *U.S. v. AT&T,* Civil Action No. 82-0192, District Court of the District of Columbia, December 17, 1982.

28. "Affidavit of Charles Brown," *U.S. v. AT&T,* Civil Action No. 82-0192, District Court of the District of Columbia, December 17, 1982.

29. "Preliminary Comments of the People of the State of California and the Public Utilities Commission of the State of California on the Plan of Reorganization," *U.S. v. AT&T,* Civil Action No. 82-0192, District Court of the District of Columbia, February 4, 1983.
30. Interview with Charles Brown, March 31, 1992.
31. Under the agreement, AT&T assumed $850 million in debt from Pacific Telephone. In addition, AT&T agreed to honor the intent of its October 1981 financial plan by converting $600 million in 1983 advances to equity. These efforts aimed to reduce Pacific's debt ratio below 47 percent at the time of divestiture with an embedded cost of debt below 9.5 percent. Virginia Dwyer to Robert DiGiorgio, April 1, 1983. See also "Motion to Approve Plan of Reorganization as Further Amended," *U.S. v. AT&T,* Civil Action No. 82-0192, District Court of the District of Columbia, April 7, 1983.

CHAPTER SIX

1. Interview with Sam Ginn, February 17, 2000.
2. John Brooks, *Telephone: The First Hundred Years* (New York, 1976), p. 205. William H. Whyte, Jr., comments on the significance of this program in *The Organization Man* (New York, 1957 ed.), p. 111. "Sammie Ginn Awarded Sloan Fellowship," *Southern Lines,* May 27, 1968, p. 1. Interview with Sam Ginn by Tracey Panek, September 27, 1999.
3. Interview with Sam Ginn, February 17, 2000. For the research report see Sam Ginn Papers, 1961–1969, Box 002:004, AirTouch Archives.
4. Eric John Abrahamson with Marjorie Wilkens, *Learning to Compete: A History of Pacific Telesis Group* (San Francisco, 1994), p. 197. SBC Communications Archives.
5. American Telephone & Telegraph Co., *Information Statement and Prospectus,* November 8, 1983.
6. Interview with Sam Ginn, February 10, 1992.
7. Joint Hearings, Subcommittee on Telecommunications, Consumer Protection, and Finance et al., House of Representatives, January 26 and 28, 1982, p. 34.
8. Cellular Telecommunications Industry Association, *Semi-Annual Wireless Industry Survey Results, June 1985 to December 2000.*
9. The FCC issued a construction permit for the cellular system in Los Angeles on March 31, 1983. See 11 CPUC 2d at 838.
10. Pacific Telesis renamed its cellular company a number of times between 1984 and 1994. For the reader's convenience, we use the name PacTel Cellular or simply PacTel throughout this period. The company began as PacTel Mobile Access in 1984. That year, the company established PacTel Mobile Services as a separate subsidiary to sell cellular telephones and equipment. It also established an international division to look for opportunities to develop cellular systems abroad. Following the completion of the acquisition of Communications Industries in 1986, PacTel Mobile Access became PacTel Cellular; together with PacTel Paging, both companies became subsidiaries of PacTel Personal Communications (PerCom). Meanwhile, in partnerships where it held a minority position, such as San Francisco, service was marketed under a variety of different brand names. California Public Utilities Commission, Public Staff Division, Telecommunications Audit Branch, "A Report on Pacific Bell's Affiliated/Subsidiary Companies," June 3, 1986, pp. 4-1 and 4-2.
11. Interview with Phil Quigley, May 7, 1992 on the issue of going to AT&T for capital. See Interview with Craig McCaw, June 8, 2000, on the issue of AT&T eliminating

cell sites. McCaw says: "one of the great heresies of the Bell System was that they cut capital for cellular in 1982, '83, and '84, so that they cut cell sites out of every city in the country, and it not only made the portable service terrible, it made the mobile service poor. And for whatever reason, just before the breakup, they cut their capital budgets, and so these markets went on the air with 40 percent fewer cell sites than they should have had."

12. 11 CPUC 2d at 838.

13. Interview with Phil Quigley, May 7, 1992.

14. The FCC had already ruled once on this issue. On April 7, 1981, the Commission concluded that the head start would not create a significant competitive advantage. But it did leave the door open for companies to ask the government to take specific action in particular markets if competitors could show that a head start would not be in the public interest. From a neo-Schumpeterian perspective, of course, a head start in innovation would always be in the public interest. Federal Communications Commission, *Report and Order of April 7, 1981,* in FCC Docket No. 79-318, fn. 57, p. 24.

15. 11 CPUC 2d at 838.

16. *Telocator Network of America Bulletin,* June 15, 1984, p. 2. With this call, Los Angeles became the eighth major city in the U.S. with cellular service. In the same month, systems were activated in Minneapolis, Baltimore, and Buffalo.

17. "Interview with Phil Quigley," *American Telecom Quarterly,* draft in Comm 010, PacTel Mobile Access, Publicity 1984–85, AirTouch Archives. Abrahamson with Wilkens, *Learning to Compete.* SBC Communications Archives.

18. Ellen Benoit, "Now You Hear It ...," *Financial World,* October 14, 1986, p. 18.

19. Dick Purcell, "Here Come the Carriers," *Communications,* January 1988.

20. California Commission on State Finance, in *California Almanac* (Santa Barbara, 1991), p. 268.

21. Nationwide, state regulators approved an unprecedented $10.5 billion in rate hikes for the Bell Operating Companies between 1982 and 1986, out of $70 billion in annual revenues. See Paul Eric Teske, *After Divestiture: The Political Economy of State Telecommunications Regulation* (Albany, 1990), p. 40. In California, the accumulated rate awards granted to Pacific by the CPUC between 1980 and 1984 exceeded $1.7 billion in additional revenues per year. Intrastate access charges imposed on AT&T virtually guaranteed another $1.2 billion a year in revenues. Pacific Bell's allowed rate of return of 17.4 percent was the highest among all the Bell Operating Companies. See California Public Utilities Commission decisions, CPUC application numbers: 59269, 59849, 60510, and 83-01-22. On access charges, see California Public Utilities Commission, *Decision 83-12-024,* December 7, 1983.

CHAPTER SEVEN

1. Interview with Arun Sarin, March 29, 2000.

2. AnnaLee Saxenian, "Silicon Valley's New Immigrant Entrepreneurs," Public Policy Institute of California, June 1999. Don Clark, "South Asian 'Angels' Reap Riches, Spread Wealth in Silicon Valley," *Wall Street Journal,* May 2, 2000, p. B1. Michael Lewis, *The New New Thing: A Silicon Valley Story* (New York, 2000), pp. 114–16.

3. Interview with Arun Sarin, June 1, 1993.

4. After the breakup, Pacific Telesis and GTE reached an agreement to consolidate their holdings. Telesis gave up most of its stake in the license in San Francisco

(keeping only 3 percent of the shares to preserve their right of first refusal if GTE chose to sell) in exchange for an equivalent value in the Los Angeles market. Interview with Phil Quigley, May 7, 1992.

5. Interview with Phil Quigley, May 7, 1992.
6. Interview with Craig McCaw, June 8, 2000. See also O. Casey Corr, *Money from Thin Air* (New York, 2000), and Quentin Hardy, "Craig's Higher Calling," *Forbes,* June 12, 2000.
7. Interview with Craig McCaw, June 8, 2000.
8. *San Francisco Examiner,* April 3, 1985, p. C1; interview with Sam Ginn, May 3, 1993.
9. Interview with Lee Cox, June 9, 1993.
10. Interview with Arun Sarin, June 1, 1993.
11. *Communications Daily,* May 23, 1985, p. 3.
12. Interview with Craig McCaw, June 8, 2000.
13. Interview with Vern Tyerman, July 13, 2000.
14. Corr, *Money from Thin Air,* p. 151.
15. Robert Ristelhueber, "Communications Industries Accepts PacTel $431M Buy Offer," *Electronic News,* May 27, 1985, p. 1.
16. Clayton Niles, unpublished autobiography, draft dated August 23, 2000.
17. Interview with Clayton Niles, August 22, 2000.
18. Interview with Arun Sarin, June 1, 1993.
19. Niles, unpublished autobiography. Interview with Sam Ginn, May 3, 1993.
20. Interview with Arun Sarin, June 1, 1993.
21. Eric John Abrahamson with Marjorie Wilkens, *Learning to Compete: A History of Pacific Telesis Group* (San Francisco, 1994), p. 258. SBC Communications Archives.
22. Interview with John Hulse, May 14, 1993.
23. Interview with Arun Sarin, June 1, 1993.
24. Charles Elmore, "Former Bell 'Brothers' Now Foes in Cellular Market," *Atlanta Constitution,* September 29, 1987, p. D1.
25. "Worrisome: Metromedia's Leveraged Buyout," *Fortune,* January 9, 1984, p. 8.
26. Julia Reed, "The Billionaire Who Just Won't Quit," *U.S. News & World Report,* June 27, 1988, p. 25.
27. James B. Murray, Jr., *Wireless Nation: The Frenzied Launch of the Cellular Revolution in America* (Cambridge, MA, 2001), pp. 192–3.
28. Barnaby J. Feder, "Taking Telesis off the Beaten Track," *New York Times,* November 20, 1988, p. F7.
29. Claire Pool, "Southwestern Belle," *Forbes,* April 16, 1990, p. 89.
30. Stuart Gannes and Lynn Fleary, "Behold, the Bell Tel Cell War; The Mobile Phone Market Is Abuzz," *Fortune,* December 22, 1986, p. 97.
31. Interview with Sam Ginn, July 20, 1993.
32. Julie Amparano, "At Pacific Telesis, Ginn Pushes Change," *Wall Street Journal,* April 25, 1988, p. 34.
33. Henry Schulman, "Telesis Buys Detroit Franchise," *Oakland Tribune,* September 29, 1987, p. 1B.
34. Elmore, "Former Bell 'Brothers' Now Foes in Cellular Market," p. D1.
35. Bruce Wasserstein, *Big Deal: The Battle for Control of America's Leading Corporations* (New York, 1997), p. 311.
36. Organization for Economic Cooperation and Development, *Telecommunication Infrastructure: The Benefits of Competition* (Paris, 1995), p. 68.

37. McCaw sold his cable assets to Washington Redskins owner Jack Kent Cooke for $755 million.
38. Corr, *Money from Thin Air.*
39. Sam Ginn to Craig McCaw, August 25, 1987, in Ginn 01 Chron Files-1, 1987, Air-Touch Archives.
40. Interview with Phil Quigley, June 10, 1993.

CHAPTER EIGHT

1. Interview with F. Craig Farrill, June 5, 2001.
2. Interview with Craig McCaw, June 8, 2000.
3. Gregory P. Allen, "The Static That Plagues Cellular Phones," *Business Week,* November 2, 1987.
4. CI had negotiated a Type-2 interconnection in San Diego, which meant that calls came into the landline network at the tandem level and were more easily routed through the network. Interview with Barry Lewis, May 17, 1993.
5. Interview with Barry Lewis, May 17, 1993. For information on the Bell System's historic approach to customers and marketing, see Peter Temin with Louis Galambos, *The Fall of the Bell System* (Cambridge, 1987), pp. 167–9, and Alvin Von Auw, *Heritage & Destiny* (New York, 1983), pp. 162–3.
6. Eric John Abrahamson with Marjorie Wilkens, *Learning to Compete: A History of Pacific Telesis Group* (San Francisco, 1994). SBC Communications Archives.
7. Interview with Barry Lewis, May 17, 1993.
8. PacTel estimated that Los Angeles would account for 85 percent of its total revenues, net income, and value through 1990. Abrahamson with Wilkens, *Learning to Compete.* SBC Communications Archives.
9. Interview with Phil Quigley, June 10, 1993.
10. Interview with Arun Sarin, June 1, 1993.
11. Interview with Craig McCaw, June 8, 2000.
12. The Cellular Telecommunications Industry Association calculated that there were 682,000 customers nationwide. The U.S. Department of Commerce estimated 700,000 customers. U.S. Department of Commerce, *1987 Industrial Outlook.*
13. Nationally, the average revenue per customer for cellular service hovered around $150, which paid for roughly five hours of calls during peak hours. Stuart Gannes and Lynn Fleary, "Behold, the Bell Tel Cell War; The Mobile Phone Market Is Abuzz," *Fortune,* December 22, 1986, p. 97.
14. At one point, when PacTel was still fighting regulatory battles for the right to turn on its service in Los Angeles in 1984, the nonwireline competitor in Los Angeles tried to counter PacTel's claim that it was trying to meet the needs of the Olympics by pointing out that PacTel's system had been designed to serve vehicles and not the kinds of portables that would be needed by Olympics organizers. See 11 CPUC 2d at 838.
15. Gretchen Morgenson, "A Pager in Every Pocket," *Forbes,* December 21, 1992, p. 210.
16. Interview with Don Sledge, May 26, 1993.

CHAPTER NINE

1. Don Sledge handwritten notes, Neels Files, Box 00052, China Files, Shanghai, Air-Touch Archives.
2. Interview with Don Sledge, September 19, 2000.

3. Don Sledge handwritten notes, Neels Files, Box 00052, China Files, Shanghai, Air-Touch Archives. Also, interview with Don Sledge, September 19, 2000.
4. Interviews with Don Sledge, May 26, 1993, and September 19, 2000.
5. We have changed Ms. Lee's name to protect her privacy.
6. Don Sledge, "Memorandum for file," August 19, 1983, AirTouch Archives.
7. Eric John Abrahamson with Marjorie Wilkens, *Learning to Compete: A History of Pacific Telesis Group* (San Francisco, 1994), p. 212. SBC Communications Archives.
8. Don Sledge handwritten notes, Neels Files, Box 00052, China Files, Shanghai, Air-Touch Archives.
9. Shanghai International Trust and Investment Corporation (SITICO) website, ⟨www.sh.cei.gov.cn⟩.
10. Interview with Don Sledge, May 26, 1993.
11. *World Almanac and Book of Facts, 1983* (New York, 1983), pp. 509, 590.
12. Klitgaard memo, November 29, 1983, China Files, AirTouch Archives.
13. Don Sledge, "Memorandum for file," August 19, 1983. Sledge also met with PA Consultants, who were already using the PacTel International name, to negotiate acquisition of the name. As part of the deal, Pacific Telesis gave PA a consulting agreement. PA later helped PacTel in a number of international markets, including Germany. Interview with Don Sledge, September 19, 2000.
14. Don Sledge, "Memorandum for file," August 19, 1983, AirTouch Archives.
15. Ibid.
16. Ibid.
17. Interview with Don Sledge, May 26, 1993, p. 16.
18. Maryellen Hessel, "Treatment, PTT/China Videotape," June 25, 1983, AirTouch Archives.
19. Interview with Don Sledge, May 26, 1993.
20. Interview with Sam Ginn, March 23, 2000.
21. Don Sledge, "Memorandum for file," August 19, 1983, p. 5. "Itinerary August 21–September 5, 1983, S.L. Ginn & D.L. Sledge," AirTouch Archives.
22. Interview with Sam Ginn, March 23, 2000.
23. Interview with Sam Ginn, May 3, 1993.
24. Interview with Sam Ginn, March 23, 2000.
25. Interview with Don Sledge, May 26, 1993.
26. Ibid.
27. Abrahamson with Wilkens, *Learning to Compete*. SBC Communications Archives.
28. Interview with Don Sledge, May 26, 1993. Interview with Robert Dalenberg, April 19, 1993.
29. Thomas J. Klitgaard, Memorandum, "Pacific Telephone International; Shanghai: Chen Xing Yan," November 29, 1983, in AirTouch Archives, Jan Neels Files, Box 00052.
30. Thomas J. Klitgaard, Memorandum, "Pacific Telephone International: Kunming: Second Conference with Mr. Lin, November 25, 1983," November 29, 1983 in Air-Touch Archives, Jan Neels Files, Box 00052.
31. *U.S. v. Western Electric Co., Inc.,* 592 F. Supp 846 (1984), p. 859.
32. *U.S. v. Western Electric Co., Inc.,* Civil Action No. 82-192, Transcript, April 11, 1984, p. 11.
33. Ibid., p. 57.
34. *U.S. v. Western Electric Co., Inc.,* 592 F. Supp 846 (1984), pp. 858–61.
35. Interview with Sam Ginn, May 3, 1993.

36. Abrahamson with Wilkens, *Learning to Compete*. SBC Communications Archives.
37. Interview with Don Sledge, May 26, 1993.
38. *San Francisco Examiner,* May 1, 1985, p. C2.

CHAPTER TEN

1. "Policy & Rules Concerning the Furnishing of Customer Premises Equipment, Enhanced Services & Cellular Communications by the Bell Operating Companies," 95 FCC 2d (1983). Cited in Michael K. Kellogg et al., *Federal Telecommunications Law* (Boston, 1992), p. 516.
2. Interview with Carl Danner, July 22, 1993.
3. In the industry, the transfer of revenues from long-distance to local-exchange operations was known as "separations." In the 1940s, the CPUC played a leading role in promoting this subsidy as national policy. For an overview of the separations issue, see Peter Temin with Louis Galambos, *The Fall of the Bell System* (Cambridge, 1987), pp. 19–27, as well as Richard Gabel, *Development of Separations Principles in the Telephone Industry* (East Lansing, MI, 1967).
4. California Public Utilities Commission, *Decision 86-01-026,* January 10, 1986. See also California Public Utilities Commission, Public Staff Division, Telecommunications Audit Branch, "Report on the Affiliated Relationships of Pacific Bell and the Pacific Telesis Group," April 22, 1985.
5. California Public Utilities Commission, "Report on the Affiliated Relationships of Pacific Bell and the Pacific Telesis Group."
6. California Public Utilities Commission, *Decision 86-01-026,* January 10, 1986.
7. The five-year plan adopted in 1983 called for the diversified businesses to contribute $97 million to corporate net income by 1988, or approximately 6 percent of Pacific Telesis earnings. See Eric John Abrahamson with Marjorie Wilkens, *Learning to Compete: A History of Pacific Telesis Group* (San Francisco, 1994), p. 200. SBC Communications Archives.
8. Barnaby J. Feder, "Taking Telesis off the Beaten Track," *New York Times,* November 20, 1988, p. F7.
9. Sam Ginn to Donald Guinn, Memorandum, October 15, 1985, Sam Ginn files.
10. Ellen Benoit, "Now You Hear It ...," *Financial World,* October 14, 1986, p. 18.
11. Angeline Pantages, "NYNEX Corp.," *Datamation,* June 15, 1992, p. 119.
12. Charles W. Ross, "Pacific Telesis Head Turns over Firm to Hand-Picked Heir," *San Diego Union,* April 23, 1988, p. E2.
13. Mireya Navarro, "New Chief Sees Growth for PacTel," *San Francisco Examiner,* July 16, 1987, p. C1.

CHAPTER ELEVEN

1. Donnie Radcliffe and Joseph McLellan, "Fond Farewells at the White House," *Washington Post,* November 17, 1988, p. C1.
2. Itinerary for Mr. Ginn, "Dinner at the White House," November 16, 1988, Ginn 003:000, AirTouch Archives.
3. Interview with Sam Ginn, March 23, 2000, pp. 61–2. Interview with Ann Ginn, March 29, 2000, pp. 29–30.
4. Eli Noam, *Telecommunications in Europe* (New York, 1992); Adrienne Hardman, "Private Lines, Public Lessons," *Financial World,* September 15, 1992, p. 64.

5. Peter Temin with Louis Galambos, *The Fall of the Bell System* (Cambridge, 1987). See also Alan Stone, *Public Service Liberalism: Telecommunications and Transitions in Public Policy* (Princeton, 1991).

6. See, for example, George Stigler, "The Theory of Economic Regulation," *Bell Journal of Economics* (Spring 1971), pp. 3–21, and Gabriel Kolko, *The Triumph of Conservatism: A Reinterpretation of American History, 1900–1916* (New York, 1963).

7. Ronald H. Coase, "The Problem of Social Cost," *Journal of Law and Economics* 3 (1930), pp. 1–24. For examples of industry studies, see Richard E. Caves, *Air Transport and Its Regulators: An Industry Study* (Cambridge, MA, 1962), and Michael E. Levine, "Is Regulation Necessary? California Air Transportation and National Regulatory Policy," *Yale Law Journal* 74, no. 8 (1965), pp. 1416–47. Alfred Kahn summarized this growing critique in his comprehensive two-volume text *The Economics of Regulation: Principles and Institutions* (New York, 1970–71).

8. Federal Communications Commission, *Second Computer Inquiry,* 77 FCC 2d 384 (1980).

9. The Carter Report to the U.K. Parliament in 1977 emphasized the inability of British Telecommunications to meet the needs of corporate users for a new generation of specialized corporate networks with value-added services. C. F. Carter, "Report of the Post Office Review Committee," presented to the U.K. Parliament by the Secretary of State for Industry, July 1977, cited in Andrew Davies, "Innovation in Large Technical Systems: The Case of Telecommunications," *Industrial and Corporate Change* 5, no. 4 (1996), pp. 1143–80.

10. Andrew Kupfer, "Ma Bell and Seven Babies Go Global," *Fortune,* November 4, 1991, p. 118.

11. Clint Bolick, "Thatcher's Revolution: Deregulation and Political Transformation," *Yale Journal on Regulation* 12 (1995), pp. 527–48.

12. Pablo T. Spiller and Ingo Vogelsgang, "The United Kingdom: A Pacesetter in Regulatory Incentives," in Brian Levy and Pablo T. Spiller (Eds.), *Regulations, Institutions, and Commitment: Comparative Studies in Telecommunications* (Cambridge, 1996), pp. 79–120. See also John Harper, *Monopoly and Competition in British Telecommunications* (London, 1997).

13. Department of Trade and Industry, *Competition and Choice: Telecommunications Policy for the 1990s* (London, 1990), p. 11. John Braithwaite and Peter Drahos, *Global Business Regulation* (Cambridge, 2000), p. 323.

14. Department of Trade and Industry, *Competition and Choice,* p. 11. See also Edgar Grande, "The New Role of the State in Telecommunications: An International Comparison," *Western European Politics,* July 1994, p. 138.

15. PacTel got its first taste of British deregulation in 1984 when John Gaulding and Vern Tyerman went to London to talk with the people at Mercury. Frustrated by their efforts to get an interconnect agreement with BT, Mercury hired Tyerman for a month to help with the negotiations. Since PacTel had not received a waiver for international business, his services were not billed for nearly six months. (Interview with Vern Tyerman, July 13, 2000.)

Around this same period, Ginn went to London in connection with PacTel's acquisition of the One 2 One voice messaging service. On this visit, Ginn was introduced to the Thatcher reforms when he met with an official from the newly created telecommunications regulatory agency, OffTel. "He was the first regulator I had ever seen," Ginn recalled, "that had internalized, and believed in, the competitive model. You didn't have to jump through any hoops. You didn't have to produce any studies. He was an enabler." (Interview with Sam Ginn, March 23, 2000, p. 60.)

16. Ernest Weiss, "Telecommunications Policy: The User's Need for Telecommunications Systems. A Review of Trends in Europe," in J. F. Rada and G. R. Pipe (Eds.), *Communication Regulation and International Business* (Amsterdam, 1984). See also Grande, "The New Role of the State in Telecommunications," p. 138. Japan also began to privatize its national phone service, Nippon Telegraph and Telephone Corporation, in 1985. David Levi-Faur, "The Competition State as a Neomercantilist State: Understanding the Restructuring of National and Global Telecommunications," *Journal of Socio-Economics,* Nov.–Dec. 1998, p. 665.

17. *Towards a Dynamic European Economy: Green Paper on the Development of the Common Market for Telecommunications Services and Equipment,* COM (87) 290, Brussels, June 30, 1987. See also Lionel H. Olmer, "Telecommunications in the Single European Market," Issue Paper, Advisory Committee on International Communications and Information Policy, July 17, 1989, p. 2, and Leonard Waverman and Esen Sirel, "European Telecommunications Markets on the Verge of Full Liberalization," *Journal of Economic Perspectives* 11, no. 4 (1997), pp. 113–26.

18. Braithwaite and Drahos, *Global Business Regulation,* p. 323.

19. Eric John Abrahamson with Marjorie Wilkens, *Learning to Compete: A History of Pacific Telesis Group* (San Francisco, 1994). SBC Communications Archives.

20. "PacTel Companies Highlights," May 1986 Management Report, AirTouch Archives.

21. AirTouch Archives, Box 081993C, India files.

22. Robert Strauss, "Nordic Rivalry," *Red Herring,* April 2000, p. 218.

23. Lawrence Harte et al., *Cellular and PCS: The Big Picture* (New York, 1997), p. 24.

24. Ibid., p. 55.

25. "Tomorrow's Bulging Pockets," *The Economist,* October 5, 1991, p. T7.

26. Raj Pandya, *Mobile and Personal Communication Services and Systems* (New York, 2000), pp. 27–8.

27. Harte, *Cellular and PCS,* p. 64.

CHAPTER TWELVE

1. Kathleen Pender, "Pac Bell's New Way to Think," *San Francisco Chronicle,* March 23, 1987, p. 1.

2. *Contra Costa Times,* June 11, 1987, p. 21A.

3. Interview with Lee Cox by Tracey Panek, January 6, 1997.

4. Eric John Abrahamson with Marjorie Wilkens, *Learning to Compete: A History of Pacific Telesis Group* (San Francisco, 1994), pp. 225–8. SBC Communications Archives.

5. Interview with Jan Neels, April 27, 2000.

6. Ginn had struggled with the issue of who should lead PacTel International. After replacing Sledge with interim president Ross Brown, he considered several possibilities; at the top of his list in January was Arun Sarin. But given the skills and experience that Neels had, Ginn selected him.

CHAPTER THIRTEEN

1. Group Seven Associates, "Market Demand Estimates for Cellular Telephone Service in the Federal Republic of Germany," February 1989, in ISCM 001 Portable Phones Study – Germany, AirTouch Archives.

2. On the role of global alliances to facilitate foreign market penetration in telecommunications, see Mark A. Jamison, "Emerging Patterns in Global Telecommunications Alliances and Mergers," *Industrial and Corporate Change* 7, no. 4 (1998), pp. 706–7.

3. Horst A. Wessel, "Mannesmann 1890: A European Enterprise with an International Perspective," *Journal of European Economic History* 29, nos. 2–3 (2000), pp. 335–56.

4. Adrienne Linsenmeyer, "If There's a Seat at the Bar ...," *Financial World,* July 23, 1991, p. 24. See also Peter Fuhrman, "Struggling to Adapt," *Forbes,* April 26, 1993, p. 102.

5. Linsenmeyer, "If There's a Seat at the Bar ...," p. 24.

6. Interview with Jan Neels, June 13, 2000.

7. Mark Jamison highlights the pattern of global alliances and mergers in telecommunications services in the mid-1990s, but he misses the role of wireless as a focus of foreign investment by the Baby Bells. He therefore sees telecommunications restructuring as a dance among telecom giants. In wireless in Europe and elsewhere, the Baby Bells helped bring into telecommunications new entrants that would become major players, including Mannesmann, Olivetti, and Compagnie Generale des Eaux SA. Jamison, "Emerging Patterns in Global Telecommunications Alliances and Mergers."

8. PacTel Development Corp., May 2, 1990, p. 19, Box RS9 010, AirTouch Archives.

9. The FCC had given Stenbeck's firm, Millicom, the third experimental cellular license in the United States. This U.S. experience proved invaluable when Stenbeck went to Britain. Daniel A. Levinthal, "The Slow Pace of Rapid Technological Change: Gradualism and Punctuation in Technological Change," *Industrial and Corporate Change* 7, no. 2 (1998), pp. 217–47.

10. Richard Grant, "AirTouch-Vodafone, Millicom Likely to Cross Paths in Emerging Markets," *Knight-Ridder/Tribune Business News,* January 25, 1989.

11. Interview with Chris Gent, June 21, 2000.

12. Interview with Jan Neels, June 13, 2000.

13. Interview with Jan Neels by Tracey Panek, December 4, 1998, p. 2.

14. Box 081993C, File German Cellular, AirTouch Archives.

15. Interview with Erhart Meixner, June 27, 2000, p. 5.

16. Craig Farrill notebooks, June 22, 1989, AirTouch Archives.

17. Interview with Sam Ginn, December 12, 2000; interview with Erhart Meixner, June 27, 2000. Eli Noam, *Telecommunications in Europe* (New York, 1992), p. 98.

18. Interview with Erhart Meixner, June 27, 2000.

19. Ibid.

20. Ibid.

21. Interview with George Schmitt, June 6, 2000.

22. Interview with William L. Keever, March 22, 2000.

23. "International Venture," 1993 Key Issues Binder, AirTouch Archives. See also Clifford Carlsen, "PacTel Explores Cellular Overseas," *San Francisco Business Times,* December 6, 1991, p. 1.

24. Interview with William Keever and Sam Ginn, March 23, 2000.

25. Calvin Sims, "Bonn Expected to Award Cellular Contract," *New York Times,* December 7, 1989, p. C1.

26. PacTel International, "Mannesmann Mobilfunk, the D2 Privat Operator, Fact Sheet," February 1993. In Comm 004. PacTel International Press Releases.

27. Deutsche Telekom, which launched its GSM service (D1) within weeks of the start of Mannesmann Mobilfunk's service, continued to provide analog service through a separate subsidiary, C-Tel.

28. Peter Fuhrman, "Struggling to Adapt," *Forbes,* April 26, 1993, p. 102.

29. Andrew Kupfer, "Ma Bell and Seven Babies Go Global," *Fortune,* November 4, 1991, p. 118. John R. McNamara, *Economics of Innovation in the Telecommunications Industry* (New York, 1991), pp. 51–3.

30. Peter Drahos and Richard A. Joseph, "Telecommunications and Investment in the Great Supranational Regulatory Game," *Telecommunications Policy* 19, no. 8 (1995), pp. 619–35.

31. Kupfer, "Ma Bell and Seven Babies Go Global."

32. McNamara, *Economics of Innovation in the Telecommunications Industry,* pp. 51–3.

33. For an overview, see David E. M. Sappington and Dennis L. Weisman, *Designing Incentive Regulation for the Telecommunications Industry* (Cambridge, MA, 1996).

34. California Public Utilities Commission, "Competition in Local Telecommunications," May 1987. Eric John Abrahamson with Marjorie Wilkens, *Learning to Compete: A History of Pacific Telesis Group* (San Francisco, 1994). SBC Communications Archives. Also, see "Fighting on the Frontiers," *The Economist,* October 5, 1991, p. T33.

35. Under Phil Quigley, Pacific Bell began an aggressive campaign to streamline the organization. Quigley reorganized to improve customer service and marketing in preparation for competition. But political turnover on the Commission and wrangling between competitors in the hearing rooms delayed final implementation of the new framework.

36. Jack Egan, "The Sweet Song of Ma Bell," *U.S. News & World Report,* January 27, 1992, p. 69.

37. Gary Slutsker, "What Should We Be?" *Forbes,* September 28, 1992, p. 132.

38. Interview with John Hulse, May 14, 1993, p. 31.

CHAPTER FOURTEEN

1. Eric John Abrahamson with Marjorie Wilkens, *Learning to Compete: A History of Pacific Telesis Group* (San Francisco, 1994). SBC Communications Archives.

2. Interview with Sam Ginn, September 6, 2000.

3. Interview with Robert Barada, May 18, 1993, p. 43.

4. Interview with Robert Smelick, January 14, 1994.

5. Abrahamson with Wilkens, *Learning to Compete.* SBC Communications Archives.

6. Interview with Sam Ginn, December 12, 2000.

7. Interview with Robert Smelick, January 14, 1994, p. 14.

8. Charles Mason, "Cellular Power to the People ...," *Telephony,* June 24, 1991, p. 36.

9. Ginn subscribed to a vision of the telecommunications industry that was sometimes called the "Negroponte Flip." The director of MIT's Media Lab, Nicholas Negroponte had predicted that most voice traffic would eventually migrate to wireless while data traffic filled the wireline network. Some researchers at this time, however, concluded that telecommunications was not a zero-sum game. In Europe, as wireless usage increased, it actually added traffic to the networks of local exchange companies, increasing the utilization of existing facilities and enhancing economies of scale. See Organization for Economic Cooperation and Development, *Telecommunication Infrastructure: The Benefits of Competition* (Paris, 1995), pp. 56–8.

10. Interview with C. Lee Cox, June 9, 1993.
11. Ibid.
12. "The Long Road to a Spin-Off," *San Francisco Chronicle,* October 20, 1993, p. C4.
13. James A. Miles and James D. Rosenfeld, "The Effect of Voluntary Spin-Off Announcements on Shareholder Wealth," *Journal of Finance* 38, no. 5 (1983), pp. 1597–1606. For a subsequent review of this literature, see Sudha Krishnaswami and Venkat Subramaniam, "Information Asymmetry, Valuation, and the Corporate Spin-Off Decision," *Journal of Financial Economics* 53, no. 1 (1999), pp. 73–112.
14. David Stadtler, Andrew Campbell, and Richard Koch, *Breakup!: How Companies Use Spin-Offs to Gain Focus and Grow Strong* (New York, 1997), pp. 47–8.
15. Sharon Reier, "Vodafone: Counting Sheep?" *Financial World,* July 21, 1992, p. 16.
16. *San Francisco Business Times,* January 22, 1993, p. 20.
17. Abrahamson with Wilkens, *Learning to Compete.* SBC Communications Archives.
18. These regulatory restraints on the pace and direction of technological innovation confirm the view of Thomas Hughes, Andrew Davies, and others that political institutions and their related interests have a powerful impact on the paths of development in new technologies. Economic considerations faced by business leaders are often framed by these political realities as well as by the economies of scale, scope, and system suggested by the innovative technologies themselves. See Thomas Hughes, *Networks of Power: Electrification in Western Society, 1880–1930* (Baltimore, 1983), and Andrew Davies, "Innovation in Large Technical Systems: The Case of Telecommunications," *Industrial and Corporate Change* 5, no. 4 (1996), pp. 1143–80.
19. Abrahamson with Wilkens, *Learning to Compete.* SBC Communications Archives.
20. Ibid.
21. Richard Doherty, "FCC Gets 'Personal,' " *Electronic Engineering Times,* December 9, 1991, p. 1.
22. Peter W. Huber et al., *The Geodesic Network II: 1993 Report on Competition in the Telephone Industry* (Washington, DC, 1992), pp. 1.10–1.11.
23. Standard & Poor's, *Industry Surveys* (New York, January 1994), vol. 2 (M–Z), p. 35.
24. Abrahamson with Wilkens, *Learning to Compete.* SBC Communications Archives.
25. Interview with Sam Ginn, September 6, 2000.
26. Interview with Robert Smelick, January 14, 1994.
27. O. Casey Corr, *Money from Thin Air* (New York, 2000); James B. Murray, Jr., *Wireless Nation: The Frenzied Launch of the Cellular Revolution in America* (Cambridge, MA, 2001), pp. 248–9.
28. By raising the possibility that AT&T might use McCaw's wireless network as a way to bypass the Baby Bell's local exchange operations, the deal sent the shares of the Baby Bells down 5–10 percent. "The Big Break," *The Economist,* November 14, 1992, p. 75.
29. Sam Ginn to Pacific Telesis Group Board of Directors, October 20, 1992.
30. Abrahamson with Wilkens, *Learning to Compete.* SBC Communications Archives.
31. Mike Mills, "AirTouch Chief's Vision Puts Firm at Forefront," *International Herald-Tribune,* January 7, 1999.

CHAPTER FIFTEEN

1. Interview with Sam Ginn, December 12, 2000.
2. *San Francisco Business Times,* February 26, 1993, p. 6; *San Francisco Chronicle,* October 20, 1993, p. C4.

3. California Public Utilities Commission, *Decision 93-11-011,* November 2, 1993, pp. 5–6.

4. California Public Utilities Commission, *I. 93-02-028,* "Opening Brief of Pacific Telesis Group," Pacific Telesis Group, July 30, 1993, pp. 47–8.

5. The CPUC's attitude toward investment reflected a tension in the historical approach to regulation in the United States, an approach that raises key issues for the Schumpeterian perspective. Traditionally, regulators in the United States viewed telecommunications as a relatively static system where innovation took place in a slow and orderly fashion, managed by the Bell System and regulators. In this environment, regulators feared gold-plating that would inflate the rate base and increase profits for the monopoly. The regulatory system, as we have seen, had problems adjusting to the stochastic shocks delivered by the sort of fundamental technical innovations that would have sparked surges in investment and returned temporarily higher profits to innovators in a competitive market. It thus left little room for lenders or investors – key players in the dynamic Schumpeterian economy – to shape the pattern of innovation. See, for example, Johannes M. Bauer, "Market Power, Innovation, and Efficiency in Telecommunications: Schumpeter Reconsidered," *Journal of Economic Issues* 31, no. 2 (1997), pp. 557–65.

6. For operating revenues, see AirTouch Communications, *1994 Annual Report,* p. 11.

7. Interview with Patricia Eckert, August 7, 2001.

8. California Public Utilities Commission, "Opening Brief of Pacific Telesis Group," pp. 63–4.

9. California Public Utilities Commission, "Proposed Decision of ALJ Wheatland," *I. 93-02-028,* September 7, 1993.

10. Ibid. Goldman, Sachs & Co. analysts Robert B. Morris and Julie E. Kennedy told reporters that they believed a ruling that required PacTel to pay up to $500 million in compensation would reduce the value of Pacific Telesis stock by about $1 or $2 per share. "Shumway, ALJ Give Different Views of PacTel Spin-Off," *Telecommunications Reports,* September 13, 1993, pp. 7–8.

11. Sam Ginn to Pacific Telesis employees, voice-mail, September 7, 1993.

12. Sam Ginn, "Telesis Customers' Rights," Letter to the Editor, *San Francisco Examiner,* September 15, 1993, p. A18.

13. "Overview: Pacific Telesis Response to the Proposed Staff ALJ Decision Regarding the Telesis Spin-Off," Pacific Telesis Group, September 15, 1993. Filed with Notice of Ex Parte Communication, *I. 93-02-028,* September 20, 1993, before the California Public Utilities Commission.

14. Pacific Bell ran into similar trouble that fall as the CPUC struggled to resolve remaining issues related to opening competition. Critics charged that a Pacific Bell employee had improperly lobbied the Commission, and after an investigation the CPUC rescinded its decision on the issue. California Public Utilities Commission, *Decision 93-10-033, Order Rescinding Decision 93-09-076,* October 6, 1993.

15. Although the Bell Atlantic–TCI deal collapsed months later after a 13-percent drop in Bell Atlantic's stock and an FCC decision to cut cable rates by 17 percent, convergence remained a media mantra. For an overview of the effect of the announced deal on the perceived value of the other Baby Bells, see Michael Hertzel et al., "Competitive Impact of Strategic Restructuring: Evidence from the Telecommunications Industry," *Industrial and Corporate Change* 10, no. 1 (2001), pp. 207–46.

16. Carla Lazzareschi, "New Spin on Spin-Off: Analysts Balk as Pacific Telesis Resists Merger Mania," *Los Angeles Times,* October 20, 1993.

17. "Shumway, ALJ Give Different Views of PacTel Spin-Off," pp. 7–8.
18. Ibid.
19. See the Omnibus Budget Reconciliation Act of 1993, Public Law No. 103-66, 107 Stat. 312, which amended section 332 of the Communications Act of 1934. For the law's effect on the cellular industry, see Derek Yeo, "Getting Wireless Carriers Wired for Less: An Argument for Federal Regulation of LEC-CMRS Interconnection Agreements," *William & Mary Law Review* 39, no. 1 (1997), pp. 229–82.
20. Interview with Patricia Eckert, August 7, 2001.
21. California Public Utilities Commission, *Decision 93-11-011,* November 2, 1993.
22. John Eckhouse, "Telesis Wins Battle to Split in Two," *San Francisco Chronicle,* November 3, 1993, p. D1.
23. Ibid., p. D2.

CHAPTER SIXTEEN

1. Mary Lu Carnevale, "Pacific Telesis Plan to Split Up Poses Challenges," *Wall Street Journal,* December 14, 1992, p. A3.
2. John Eckhouse, "Big PacTel Spin-Off May Wait Until '94," *San Francisco Chronicle,* October 20, 1993, p. C1.
3. Interview with Barbara Riker, May 24, 2000.
4. AirTouch Communications, "Mid-Year 1994 Investor Fact Book," p. 11. PacTel Corporation, "Prospectus," December 2, 1993, p. 8.
5. AirTouch Communications, "Mid-Year 1994 Investor Fact Book," p. 39.
6. These were not the only shares Ginn received. Given his holdings in Pacific Telesis, he (like other shareholders) received a proportional distribution of the wireless company's stock amounting to 9,883 shares. As of September 30, 1993, he also held exercisable options on another 420,122 shares. PacTel Corporation, "Prospectus," December 2, 1993, p. 8.
7. "The Road Show: Behind the Scenes," *Enterprise,* January 1994.
8. Interview with Sam Ginn, April 20, 2000.
9. Interview with Barbara Riker, May 24, 2000.
10. Interview with Sam Ginn, December 12, 2000. Interview with Amy Damianakes, April 28, 2000.
11. "PacTel Slated for New Identity," *PacTel Enterprise,* August 1993, pp. 3–4.
12. Interview with Sam Ginn, April 20, 2000.
13. "PacTel Corporation Changes Name to Become AirTouch Communications," Press Release, February 16, 1994, AirTouch Archives.
14. Leslie Cauley, "AirTouch Begins Independent Life with Hefty Assets and High Hopes," *Wall Street Journal,* April 4, 1994, p. A5.
15. Stuart Elliott, "Advertising," *New York Times,* April 11, 1994.
16. Interview with Sam Ginn, April 20, 2000.
17. Sam Ginn, "How We Do Work Around Here," 1994, Comm 016:002. AirTouch Archives.
18. Interview with Amy Damianakes, April 28, 2000.
19. Interview with Mark Hickey, January 9, 2001.
20. Interview with Amy Damianakes, April 28, 2000.
21. Interview with Dwight Jasmann, May 2, 2000.
22. PacTel Corp., "Comparative Performance Measures in the Cellular Industry: Q2 1993," August 19, 1993. AirTouch Archives.

23. Interviews with Amy Damianakes, April 28, 2000; Chris Gent, June 21, 2000; Annelie Green, June 19, 2000.

CHAPTER SEVENTEEN

1. Heidi Kargman, "Land Mobile Communications: The Historical Roots," in Raymond Bowers et al. (Eds.), *Communications for a Mobile Society: An Assessment of New Technology* (Beverly Hills, 1977), p. 28. Concerns about higher accident rates due to cell phone distractions also sparked continued studies by medical and safety researchers. A much-quoted report in the *New England Journal of Medicine* found that drivers using cell phones increased their risk of a collision by a factor of four. See Donald A. Redelmeier and Robert J. Tibshirani, "Association between Cellular-Telephone Calls and Motor Vehicle Collisions," *New England Journal of Medicine* 336, no. 7 (1997), pp. 453–8. Although most people believe cell-phone use increases the risk of an accident, a study by the Insurance Research Council found that most users were not willing to give up driving and talking to diminish these risks. See Insurance Research Council, *Public Attitude Monitor* (1997), p. 3.

2. Garry A. Garrard, *Cellular Communications: Worldwide Market Development* (Boston, 1998), pp. 482–4. Further studies published after 1992 continued to find no link between cell phones and cancer. See, for example, P. D. Inskip et al., "Cellular-Telephone Use and Brain Tumors," *New England Journal of Medicine* 344, no. 2 (2001), pp. 79–86. But as health concerns persisted, government regulators in the United Kingdom and the United States promulgated standards for cell-phone radiation levels known as specific absorption rates (SARs), and the CTIA in August 2000 began requiring domestic cell-phone manufacturers to publish SAR levels on the packaging for individual phones. Nicole Harris and Scott Hensley, "Cell-Phone Makers Agree to Develop Risk Standard," *Wall Street Journal*, August 29, 2000, p. B4.

3. Jerry Hausman, "Cellular Telephone, New Products and the CPI," *NBER Working Paper Series*, National Bureau of Economic Research, Working Paper 5982 (March 1997).

4. L. F. Rakow and V. Navarro, "Remote Mothering and the Parallel Shift: Women Meet the Cellular Telephone," *Critical Studies in Mass Communication* 10 (1993), pp. 144–57.

5. Tamar Katriel, "Rethinking the Terms of Social Interaction," *Research on Language and Social Interaction* 32, nos. 1, 2 (1999), pp. 95–101.

6. California Public Utilities Commission, *Decision 90-06-025*, June 6, 1990, p. 8.

7. Ibid.

8. GAO, "Concerns about Competition in the Cellular Telephone Service Industry," July 1992, p. 19. Taking inflation into account, the average cost of service had actually fallen by about 27 percent.

9. Competition and cooperation could take many forms. Ginn, for example, had to get McCaw to stop taking employees from the Bay Area joint venture and bringing them into his own organization. Sam Ginn to Craig McCaw, February 24, 1987, Ginn 01-Chron Files-3, 1987. AirTouch Archives.

10. Michael K. Kellogg et al., *Federal Communications Law* (Boston, 1992), p. 671.

11. For an analysis of the duopoly era in U.S. cellular, see H. S. Fullerton, "Duopoly and Competition," *Telecommunications Policy* 22, no. 7 (1998), pp. 593–607.

12. "The Future, Sort of: Telecoms Alliances," *The Economist,* October 29, 1994, p. 74.

13. Harry A. Jessell, "PCS Applicants Span Media Spectrum," *Broadcasting & Cable,* November 7, 1994, p. 52.

14. Joe Flint, "MSO Seeks Set-aside for PCS License," *Broadcasting,* November 16, 1992, p. 66.

15. Deena Karadsheh, "The Wireless Primer: An Introduction to the U.S. Wireless Communications Industry," Research Group, AirTouch Communications, April 1998, p. 25.

16. David Ellen, "The Phone Flushaway: Uncle Sam Wants You to Win $20 Billion," *The New Republic,* October 9, 1989, p. 13.

17. Leo Herzel, a law student, was the first to call for spectrum auctions in 1951. Leo Herzel, "Public Interest and the Market in Color Television Regulation," *University of Chicago Law Review* 18 (1951), pp. 802–16. See also Ronald Coase, "The Federal Communications Commission," *Journal of Law and Economics* 2 (October 1959), pp. 1–40. For a discussion of Coase's approach and the search for a "middle way" in the 1960s, see Harvey J. Levin, "Spectrum Allocation without Market," *American Economic Review* 60, no. 2 (1970), pp. 209–19. For a history of the effort to adopt an auction policy, see Thomas W. Hazlett, "Assigning Property Rights to Radio Spectrum Users: Why Did FCC License Auctions Take 67 Years?" *Journal of Law and Economics* 41, no. 2 (1998), pp. 529–75. This paper was presented at the Law and Economics of Property Rights to Radio Spectrum Conference, the proceedings of which are collected in this same issue.

18. Public Law No. 103-66, 107 Stat. 312. Cited in Derek Yo, "Getting Wireless Carriers Wired for Less: An Argument for Federal Regulation of LEC-CMRS Interconnection Agreements," *William & Mary Law Review,* October 1997, pp. 229–82.

19. Warren Cohen, "Halting the Air Raid," *Washington Monthly,* June 1995, p. 30. Edward Warner, "FCC Rejects Some Proposed Reallocations, Suggests Others," *FCC Report,* August 25, 1994, p. 9.

20. Rita Koselka, "Playing Poker with Craig McCaw," *Forbes,* July 3, 1995, p. 62.

21. John von Neumann and Oskar Morgenstern, *The Theory of Games and Economic Behavior* (Princeton, 1944). The evolution of game theory is described in Robert J. Leonard, "From Parlor Games to Social Science: Von Neumann, Morgenstern and the Creation of Game Theory, 1928–1944," *Journal of Economic Literature* 33, no. 2 (1995), pp. 730–61.

22. For a recent review of the applications of game theory, see Robert Gibbons, "An Introduction to Applicable Game Theory," *Journal of Economic Perspectives* 11, no. 1 (1997), pp. 127–49.

23. Peter C. Cramton, "The FCC Spectrum Auctions: An Early Assessment," *Journal of Economics and Management Strategy* 6, no. 3 (1997), pp. 431–95, reprinted in Donald L. Alexander, *Telecommunications Policy: Have Regulators Dialed the Wrong Number* (Westport, CT, 1997), pp. 75–130.

24. Rob Norton, "Winning the Game of Business," *Fortune,* February 6, 1995, p. 36. Also, R. Preston McAfee and John McMillan, "Analyzing the Airwaves Auction," ⟨www.ssc.upenn.edu/econ/econ135/mcafeeauction.htm⟩.

25. Andrew Kupfer, "Telcos and Cable Make Dates for an Auction," *Fortune,* November 28, 1994, p. 16.

26. Jessell, "PCS Applicants Span Media Spectrum," p. 52.

27. Kimberly Patch, "Wireless Arena in Volatile State," *PC Week,* March 21, 1994, p. 134.

28. Peter Fletcher, "U.S. West Completes Multimedia 'Template,'" *Electronics,* June 13, 1994, p. 13.
29. AirTouch, "Mid-Year 1994 Investor Fact Book," p. 1.
30. Kupfer, "Telcos and Cable Make Dates for an Auction," p. 16.
31. Tom Dellecave, Jr., "McCaw to Nextel's Rescue," *Information Week,* April 24, 1995, p. 81; *Fortune,* "Craig McCaw's New Move," November 13, 1995, p. 180.
32. Michael Moeller, "Sprint, Nynex, Bell Discuss Cellular Merger," *PC Week,* September 19, 1994, p. 131.
33. Ellis Booker, "MCI, Sprint Still Waffling on Wireless," *Computerworld,* October 17, 1994, p. 7. Interview with Arun Sarin, August 3, 2001.
34. Interview with Sam Ginn, May 1, 2001.
35. Compared to AT&T–McCaw's 70 million. Michael Moeller, "Cellular Alliance to Create National Network," *PC Week,* October 24, 1994, p. 3.
36. Interview with George Schmitt, March 6, 2001.
37. Ibid. Interview with Sam Ginn, May 1, 2001.
38. "Justice Department Investigates PCS Alliances," *WTOnline,* November 10, 1994.
39. Cramton, "The FCC Spectrum Auctions," p. 94.
40. Koselka, "Playing Poker with Craig McCaw."
41. Indeed, McCaw competed with AT&T in only one market: Buffalo, New York. Cramton, "The FCC Spectrum Auctions," p. 94.
42. For an overview of one firm's bidding strategies, see David J. Salant, "Up in the Air: GTE's Experience in the MTA Auction for Personal Communication Services Licenses," *Journal of Economics & Management Strategy* 6, no. 3 (1997), pp. 549–72.
43. Interview with George Schmitt, March 6, 2001.
44. Interview with Craig McCaw, June 8, 2000.
45. *New York Times,* March 27, 1995, p. C9.
46. McAfee and McMillan, "Analyzing the Airwaves Auction."
47. O. Casey Corr, *Money from Thin Air* (New York, 2000), p. 233.
48. Cramton, "The FCC Spectrum Auctions," p. 102.
49. Koselka, "Playing Poker with Craig McCaw."
50. Cramton, "The FCC Spectrum Auctions," pp. 98–9.
51. Dan O'Shea, "PCS Auction Survivors Await Next Battle," *Telephony,* March 20, 1995, p. 6.
52. Koselka, "Playing Poker with Craig McCaw."
53. Mark Berniker, "Cable-Sprint Venture Cops 29 PCS Licenses," *Broadcasting & Cable,* March 20, 1995, p. 32.
54. Interview with Jan Neels by Tracey Panek, November 4, 1998. Interview with Arun Sarin, November 29, 2000.

CHAPTER EIGHTEEN

1. For an overview of the relationship between standards setting and competition, see Stanley M. Besen and Joseph Farrell, "Choosing How to Compete: Strategies and Tactics in Standardization," *Journal of Economic Perspectives* 8, no. 2 (1994), pp. 117–31.
2. Interview with Craig Farrill, June 5, 2001.
3. AirTouch Communications, "Mid-Year 1994 Investor Fact Book," p. 15.
4. Ann Lindstrom et al., "Making the Transition to the New Network," *Telephony,* February 25, 1991, p. S18.

5. Interview with Craig Farrill, June 5, 2001.
6. Lawrence Harte et al., *Cellular and PCS: The Big Picture* (New York, 1997), p. 52.
7. These figures were hotly debated in the industry. See Robert Poe, "Seizing the Wireless Future," *Upside,* June 2000, pp. 108–19.
8. Harte, *Cellular and PCS,* p. 127.
9. Thomas Lanning, "CTIA Backs Push for TDMA as Digital Cellular Standard," *Telephony,* January 21, 1991, p. 8.
10. Jim House, "Qualcomm, Inc.: Hold the Phone," *California Business,* June 1991, p. 11.
11. Bradley J. Fikes, "Critics Hung Up on Qualcomm's Fiscal Connections to Baby Bells," *San Diego Business Journal,* August 10, 1992, p. 6.
12. Lanning, "CTIA Backs Push for TDMA as Digital Cellular Standard," p. 8.
13. Ibid.
14. Interview with Craig Farrill, June 5, 2001.
15. George Gilder, *Telecosm: How Infinite Bandwidth Will Revolutionize Our World* (New York, 2000), p. 89.
16. Interview with Craig Farrill, June 5, 2001. Interview with Keith Kaczmarek, February 23, 2001.
17. Pressure from the United States to open the Korean telecommunications market played a role in the introduction of competition in Korea. It also helped to shine a bright light on the license allocation process. After the license was initially awarded to the Sunkyong group in 1992, a public outcry erupted when it was revealed that President Roh Tae Woo was related to the Sunkyong group chairman. Sunkyong withdrew, and a new selection process took place in 1993. AirTouch and POSCO emerged as victors in this competition. James F. Larson, *Telecommunications Revolution in Korea* (Oxford, 1995), pp. 66–71.
18. AirTouch Communications, "Mid-Year 1994 Investor Fact Book," p. 15.
19. Raj Pandya, *Mobile and Personal Communication Services and Systems* (New York, 2000), p. 73.
20. Jason Meyers, "CDMA Waits in the Wings," *Telephony,* September 4, 1995, p. 6.
21. Quentin Hardy, "An Inventor's Promise Has Companies Taking Big Cellular Gamble," *Wall Street Journal,* September 6, 1996, p. A1.
22. "Newsbriefs," *PacTel Enterprise,* May 1993, p. 1.
23. To increase capacity after their initial deployment, many GSM operators switched to half-rate (8-kb) voice-coding schemes to essentially double the number of subscribers they could serve. But this move compromised voice quality. Bruce Egan, "The Role of Wireless Communications in the Global Information Infrastructure," *Telecommunications Policy* 21, no. 5 (1997), pp. 357–85.
24. Reinhardt Krause, "PCS Competition Heats Up," *Electronic News,* November 27, 1995, p. 1.
25. Lawrence J. Curran, "PCS Will Scramble Cellular Standards," *EDN,* March 30, 1995, p. S20.
26. Hardy, "An Inventor's Promise Has Companies Taking Big Cellular Gamble," p. A1.
27. Bill Machrone, "Portable Telephones for Everyone," *PC Magazine,* July 1995, p. 83. Interview with Craig Farrill, June 5, 2001.
28. Interview with Arun Sarin, August 3, 2001.
29. "Cellular Group Expected to Back CDMA Standard," *PC Week,* June 12, 1995, p. 33.
30. Interview with Keith Kaczmarek, February 23, 2001. Interview with Craig Farrill, June 5, 2001.

31. Sam Ginn, "Employee Meeting," May 15, 1995. AirTouch Archives.
32. AirTouch Communications, "1995 Annual Report," p. 2. Customers here are actually "proportionate" customers, reflecting a pro-rated share of joint venture operations in various markets.
33. AirTouch Communications, "1995 Fact Book," p. 4.
34. Mark Landler, "Cellular Spinoff Finds Success Overseas," *New York Times,* May 6, 1996, p. C3.
35. Interview with Michael Miron, May 5, 2000.

CHAPTER NINETEEN

1. Richard H. K. Vietor, *Contrived Competition: Regulation and Deregulation in America* (Cambridge, MA, 1994), pp. 176–85.
2. For an overview of AT&T's efforts to rewrite the Communications Act in the 1970s as part of its strategy to respond to pro-competitive initiatives by the courts and administrative agencies of the federal government, see Peter Temin with Louis Galambos, *The Fall of the Bell System* (Cambridge, 1988).
3. See the discussion of these issues in Julian E. Zelizer, *Taxing America: Wilbur D. Mills, Congress, and the State, 1945–1975* (New York, 1998).
4. *Telecommunications Act* (P.L. 104-104). For various analyses, see Peter W. Huber et al., *The Telecommunications Act of 1996* (Boston, 1996), and Timothy J. Brennan, "Making Economic Sense of the Telecommunications Act of 1996," *Industrial and Corporate Change 5*, no. 4 (1996), pp. 941–61.
5. Leslie Cauley, "Cellular-Phone Spinoff Marked Start of Slide Leading to PacTel Deal," *Wall Street Journal,* April 2, 1996, p. 1; Mark Landler, "Cellular Spinoff Finds Success Overseas," *New York Times,* May 6, 1996, p. C3.
6. Jeff Peline, "Baby Bell Deals with Competition, Spin-Off," *San Francisco Chronicle,* December 6, 1995, p. B1.
7. Cauley, "Cellular-Phone Spinoff Marked Start of Slide Leading to PacTel Deal," p. 1.
8. Interview with Amy Damianakes, April 28, 2000.
9. At this time AirTouch stock was still depressed. Within a year, the market value would double. See also the annotated response to Cauley, "Cellular-Phone Spinoff Marked Start of Slide Leading to PacTel Deal." AirTouch Archives.
10. Tom Steinert-Threlkeld, "Ginn Knew What He Was Doing," *Interactive Week,* April 22, 1996, p. 23.
11. The consolidation would continue with SBC's acquisition in 1999 of Ameritech, the Chicago-based Baby Bell, for approximately $72 billion.
12. Interview with Mohan Gyani, September 8, 2000.
13. Mark A. Jamison, "Emerging Patterns in Global Telecommunications Alliances and Mergers," *Industrial and Corporate Change 7*, no. 4 (1998), pp. 695–713.
14. Testimony of Sam Ginn in *Trade Implication of Foreign Ownership Restrictions on Telecommunications Companies: Hearings Before the Subcommittee on Commerce, Trade and Hazardous Materials of the House Commerce Committee,* 104th Congress, 1st Session 35 (1995).
15. J. Gregory Sidak, *Foreign Investment in American Telecommunications* (Chicago, 1997), p. 234. Steven Globerman, "Foreign Ownership in Telecommunications: A Policy Perspective," *Telecommunications Policy 19*, no. 1 (1995), pp. 21–8.
16. Interview with Arun Sarin, August 3, 2001.

17. PacTel Cellular, "Customer Satisfaction Measurement System," September 9, 1993. AirTouch Archives.

18. Interview with Mohan Gyani, September 8, 2000.

19. For political and regulatory reasons, AT&T allowed the Bell Operating Companies to maintain their own boards of directors during the monopoly era. At Pacific Telephone, however, the board exercised some legitimate independence because of the presence of a small percentage of minority shareholders. Eric John Abrahamson with Marjorie Wilkens, *Learning to Compete: A History of Pacific Telesis Group* (San Francisco, 1994). SBC Communications Archives.

20. Interview with Mohan Gyani, September 8, 2000.

21. "Sprint Turns On Its PCS Network, Competition for San Diego Heats Up," *San Diego Union Tribune,* December 27, 1996.

22. Jonathan Marshall, "AirTouch's Latest Coup," *San Francisco Chronicle,* November 13, 1996, p. B1; AirTouch Communications, Press Release, November 12, 1996, AirTouch Archives.

23. "Wireless Industry Struggles to Link the Masses," *San Diego Union Tribune,* May 29, 1997.

24. Matt Moffett, "Cellular-Phone Boom Lures Foreign Firms to Brazilian Sell-Off," *Wall Street Journal,* March 31, 1997, p. A1.

25. Interview with Sam Ginn, April 20, 2000.

26. Interview with Michael Miron, May 5, 2000.

27. Shalom Manova et al., "Mass Market Entry of Wireless Telecommunications," *Industrial and Corporate Change* 7, no. 4 (1998), p. 681.

28. Sam Paltridge, "Upwardly Mobile Telephony," *OECD Observer* 196, Oct.–Nov., pp. 14–18.

29. Strategis Group, "US PCS Marketplace: 1995" (Strategis Group, 1995).

30. For an overview of the changes wireless companies needed to make to lower cost structures and prices and so enable mass market growth in wireless, see Manova, "Mass Market Entry of Wireless Telecommunications," pp. 679–94.

31. Arun Sarin, "Opening Remarks: Customer Loyalty Summit VIII," September 10, 1997, AirTouch Archives.

32. Researchers have noted different patterns in the development of competitive cellular markets. According to the OECD in 1995, competition in newly opened markets initially focused primarily on price differentiation. Philipp M. Nattermann, however, has noted that competition in the German cellular market evolved in three phases: from product differentiation to advertising differentiation and, finally, to price differentiation. Paltridge, "Upwardly Mobile Telephony," pp. 14–18; Philipp M. Nattermann, "The German Cellular Market: A Case of Involuntary Competition?" *Info* 1, no. 4 (1999), pp. 355–65. See also Torsten J. Gerpott et al., "Customer Retention, Loyalty, and Satisfaction in the German Mobile Cellular Telecommunications Market," *Telecommunications Policy* 25 (2001), pp. 249–69, and Tommaso M. Valletti and Martin Cave, "Competition in UK Mobile Communications," *Telecommunications Policy* 22, no. 2 (1998), pp. 109–31.

33. Interview with Arun Sarin, November 29, 2000.

34. Paltridge, "Upwardly Mobile Telephony."

35. Judith Messina, "Bell Atlantic Looks Overseas: Purchase Would Boost Wireless, but Deal May Be Out of Reach," *Crain's New York Business,* January 11, 1999, p. 3; Andrew Kupfer, "The Heir Apparent at Bell Atlantic Speaks Out," *Fortune,* November 10, 1997, p. 228.

36. Interview with Sam Ginn, February 17, 2000.
37. Interview with George Schmitt, March 6, 2001.
38. Jeff May, "New Bell Atlantic Chief Has Hands Full for 1999," *Knight-Ridder/ Tribune Business News,* January 3, 1999.
39. Lawrence T. Babbio biography, Bell Atlantic Corporation, ⟨www.ba.com⟩.
40. Interview with Sam Ginn, May 1, 2001.
41. Interview with Sam Ginn, February 17, 2000; interview with Mohan Gyani, September 8, 2000.

CHAPTER TWENTY

1. Nicholas Negroponte, "Digital Networks," *Telecommunications* 11, no. 4 (1993).
2. Joseph N. Pelton, *Wireless and Satellite Telecommunications: The Technology, the Markets & the Regulations* (Upper Saddle River, NJ, 1995), p. 24.
3. By the end of 1998, GSM led the world with more than 100 million subscribers; CDMA had 19 million, and TDMA had 18.5 million. CDMA's growth rate was remarkable. Much of the growth had taken place in Asia, with South Korea accounting for a predominant share of Asia's 11 million CDMA customers. New CDMA systems were being launched in Latin America. ("CDMA Outpaces Subscriber Growth Estimate, Noses Ahead of TDMA," *Communications Today,* September 24, 1998.) The pace of CDMA growth continued to accelerate in 1999; by mid-year, there were approximately 33 million CDMA customers worldwide. But CDMA's market acceptance seemed dwarfed by GSM's presence, which included 160 million customers worldwide by April. ("CDMA Subscriber Base Seen Hitting 33 Million," *Wireless Today,* June 15, 1999.)
4. Academic evaluations of the relationship between telecommunications and trade policy have tended to focus on telecommunications equipment markets. Deprived of a manufacturing base by the terms of the 1982 Consent Decree, the Baby Bells had to focus primarily on services in their globalization efforts. As they fought to win licenses, they enlisted the support of U.S. diplomats and trade negotiators with varying success. See Kenneth G. Robinson, Jr., et al., "Issues of International Trade," in Barry G. Cole (Ed.), *After the Breakup: Assessing the New Post-AT&T Divestiture Era* (New York, 1991), pp. 428–72. See also David Held et al., *Global Transformations: Politics, Economics and Culture* (Stanford, 1999), pp. 58–62.
5. Interview with Sam Ginn, June 7, 2001.
6. Corporate Strategy, "Updating and Refining AirTouch's Corporate Strategy," September 4, 1997, AirTouch Archives. Interview with Mike Miron, May 5, 2000.
7. Amy Damianakes to Sam Ginn et al., Memorandum, October 21, 1997, AirTouch Archives.
8. Alan Crane, "AirTouch European Assets Offer Vodafone Rich Picking," *Financial Times,* January 15, 1999, p. 22.
9. AirTouch Communications, "AirTouch Fact Book – Mid-Year 1998," p. 6.

CHAPTER TWENTY-ONE

1. "Mr. Boring Goes Shopping," *The Economist,* November 27, 1999, p. 74.
2. Alan Cowell, "Affinity for Hardball as Well as Cricket," *New York Times,* January 19, 1999.

3. Ibid.; Alan Cowell, "Vodafone's Chief Executive Has an Empire on Which to Build," *New York Times,* November 29, 1999, p. C2; "The Live Wire of Wireless," *Business Week,* June 12, 2000, p. 80.

4. "Vodafone Calling," *Business Week,* January 25, 1999, p. 24; interview with Mike Caldwell, June 20, 2000.

5. Richard Grant, "AirTouch-Vodafone, Millicom Likely to Cross Paths in Emerging Markets," *Knight-Ridder/Tribune Business News,* January 25, 1989; interview with Mike Caldwell, June 20, 2000.

6. Interview with Chris Gent, June 21, 2000.

7. Interview with Mike Caldwell, June 20, 2000.

8. Ibid.

9. Competition in the duopoly era in the United Kingdom focused on service quality rather than prices. Tommaso M. Valletti and Martin Cave, "Competition in UK Mobile Communications," *Telecommunications Policy* 22, no. 2 (1998), p. 127.

10. Early in 1987, Ginn wrote to Harrison suggesting that they should meet to share insights regarding the emerging cellular industry. See Sam Ginn to Ernest Harrison, January 28, 1987 in Ginn 01, Chron Files-3, 1987. AirTouch Archives.

11. Alyssa A. Lappen, "Making a Killing," *Forbes,* June 13, 1988, p. 8; Chris Blackhurst, "The Best Advice Money Can Buy," *Management Today,* October 1997, p. 26.

12. David Sadtler, Andrew Campbell, and Richard Koch, *Breakup!: How Companies Use Spin-Offs to Gain Focus and Grow Strong* (New York, 1997), pp. 47–8.

13. In the early days of cellular, most PTTs continued to underestimate the demand for wireless. "The main reason was a business culture derived from an era when telecommunications was not a demand-led industry." The PTTs were simultaneously investing in other services and seemed to fear (as McCaw had predicted) that wireless would cannibalize wireline revenues. With demand exceeding projections and capacity nearly full, the PTTs' marketing efforts were restrained and, behaving just as a neoclassical economist would have predicted, they made little effort to increase usage. Sam Paltridge, "Upwardly Mobile Telephony," *OECD Observer* 196 (Oct.–Nov. 1995), p. 16.

14. Interview with Mike Caldwell, June 20, 2000.

15. Interview with Chris Gent, June 21, 2000.

16. As the OECD asserted, this "skimming the cream" strategy was typical of the duopolist's approach to the cellular market; it involved, among other things, focusing on business users and ignoring consumers. The social costs of such a strategy were significant and made the case for policymakers who favored more competition. Paltridge, "Upwardly Mobile Telephony," p. 16.

17. Interview with Chris Gent, June 21, 2000. During the duopoly period in the U.K. market, prices were remarkably stable and neither company did much to differentiate customers or service options. With the imminent introduction of PCN competitors in 1993, both Vodafone and Cellnet adopted consumer-oriented product differentiation plans in an effort to lock in customers before the competition arrived. Competition, however, had little effect on prices through mid-1996. Valletti and Cave, "Competition in UK Mobile Communications," pp. 109–31.

18. Paltridge, "Upwardly Mobile Telephony," pp. 14–18.

19. Valletti and Cave, "Competition in UK Mobile Communications," p. 115.

20. Interview with Chris Gent, June 21, 2000.

21. Vodafone Plc, *1999 Annual Report,* p. 7; interview with Mike Caldwell, June 20, 2000.

22. Interview with Mike Caldwell, June 20, 2000.
23. Vodafone owned a 17.2-percent stake in PCS operator E-Plus Mobilfunk, the third digital mobile competitor. Launched in mid-1994, E-Plus struggled to compete with the incumbents – Mannesmann Mobilfunk (D2) and Deutsche Telekom (D1) – despite what was arguably a technically superior network. By January 1998, E-Plus had gained less than 15 percent of the German mobile market (roughly 1 million subscribers). Its growth had come primarily at the expense of Deutsche Telekom's increasingly irrelevant analog subsidiary C-Tel. Philipp M. Nattermann, "The German Cellular Market: A Case of Involuntary Competition?" *Info* 1, no. 4 (1999), pp. 355–65. See also Vodafone AirTouch, "Listing Particulars Relating to the Issue of Ordinary Shares in Vodafone AirTouch Plc in Connection with the Offer for Mannesmann AG."
24. Interview with Chris Gent, June 21, 2000.
25. Ibid.
26. Vodafone Plc and AirTouch Communications, *Proxy Statement/Prospectus: The Merger of Vodafone and AirTouch,* April 22, 1999, p. 29; interview with Chris Gent, June 21, 2000.
27. Date provided by Nancy Cobb, Executive Assistant to Chris Gent, March 20, 2001.
28. Interview with Chris Gent, June 21, 2000.
29. Ibid.
30. Ibid. Alan Cowell, "Affinity for Hardball as Well as Cricket."
31. Interview with Arun Sarin, August 3, 2001.
32. Interview with Chris Gent, June 21, 2000.
33. Ibid.
34. Ibid.
35. Ibid.
36. Interview with Mohan Gyani, September 8, 2000. Interview with Sam Ginn, May 1, 2001.
37. At the end of 1995, for example, Vodafone had slightly fewer customers than AirTouch but nearly 37 percent more revenues and almost 370 percent higher net income. See Organization for Economic Cooperation and Development, *Communications Outlook 1997* (Paris, 1997), vol. 1, p. 13. In making its offer, Vodafone banked on a buoyant stock market and the promise that it would be able to derive enough efficiencies from the deal to avoid dilution for its shareholders. WorldCom had used a similar strategy to become one of the twenty largest companies in the world following a series of acquisitions culminating with that of MCI (announced September 1997). For background on the role of stock swaps in the consolidation of the global telecommunications industry, see Peter Curwen, "Telcos Take Over the World," *Info* 1, no. 3 (1999), pp. 239–51.
38. Peter Curwen, "Mobility Rules? Impacts of the Vodafone AirTouch–Mannesmann Take-Over," *Info* 2, no. 1 (February 2000), pp. 25–40. Interview with Chris Gent, June 21, 2000.
39. Interview with Mohan Gyani, September 8, 2000.
40. Interview with Sam Ginn, May 1, 2001.
41. Ibid.
42. Interview with Chris Gent, June 21, 2000. Interview with Sam Ginn, May 1, 2001. Also, "Travel Itinerary, Egypt, Sweden, Germany, Italy, England, October 20–November 8, 1998," in Sam Ginn personal files.

43. Joan Nix and David Gabel, "AT&T's Strategic Response to Competition: Why Not Preempt Entry?" *Journal of Economic History* 53, no. 2 (1993), pp. 377–87.
44. Rebecca Blumenstein, "On Eve of IPO Hesse to Leave AT&T Wireless," *Wall Street Journal,* March 9, 2000, p. B1.
45. Robert A. Steuernagel, *Wireless Marketing* (New York, 1999), p. 4.
46. Interview with Mohan Gyani, September 8, 2000.

CHAPTER TWENTY-TWO

1. Elizabeth Douglass, "Giant Bell Atlantic May Be in Talks to Acquire AirTouch," *Los Angeles Times,* January 1, 1999, p. C1:4. Laura M. Holson, "In Courtship of AirTouch, Heavy Dates and Hefty Stakes," *New York Times,* January 18, 1999, p. C1. Janet Guyon, "Dial 'T' for Trouble," *Fortune,* February 15, 1999, p. 28. Interview with Mohan Gyani, September 8, 2000. Interview with Craig Jorgens, May 23, 2001. Interview with Sam Ginn, February 17, 2000.
2. Interview with Sam Ginn, February 17, 2000.
3. Holson, "In Courtship of AirTouch, Heavy Dates and Hefty Stakes," p. C6.
4. Interview with Chris Gent, June 21, 2000.
5. Simon Wilde, "Gough Hat-Trick Boosts England," *London Times,* January 3, 1999, ⟨www.Sunday-times.co.uk/news/pages/sti/99/01/03/sticricri01002.html⟩.
6. Interview with Chris Gent, June 21, 2000.
7. Michael E. Kanell, "British Firm Tops Bell Atlantic's Offer for AirTouch Communications," *Knight-Ridder/Tribune Business News,* January 7, 1999; see also *New York Times,* January 6, 1999, p. C1. Holson, "In Courtship of AirTouch, Heavy Dates and Hefty Stakes," p. C6. Vodafone/AirTouch, "Prospectus/Proxy, Merger of Vodafone and AirTouch," p. 30.
8. Holson, "In Courtship of AirTouch, Heavy Dates and Hefty Stakes," p. C6. Interview with Sam Ginn, February 17, 2000.
9. Interview with Sam Ginn, February 17, 2000.
10. Mark Tran, "MCI Tries to Foil Bid for AirTouch," *Guardian,* January 8, 1999, Sec. 1, p. 21:7. Holson, "In Courtship of AirTouch, Heavy Dates and Hefty Stakes," p. C6. Interview with Sam Ginn, February 17, 2000.
11. Judith Messina, "Bell Atlantic Looks Overseas: Purchase Would Boost Wireless, but Deal May Be Out of Reach," *Crain's New York Business,* January 11, 1999, p. 3.
12. Interview with Chris Gent, June 21, 2000.
13. "Bell Atlantic Clears Up One AirTouch Accounting Issue," *Wireless NOW,* January 15, 1999, from story by Reuters, AirTouch Archives.
14. "AirTouch Shareholders Look for Legal Best Deal," *Wireless NOW,* January 7, 1999, AirTouch Archives.
15. Vodafone/AirTouch, "Prospectus/Proxy, Merger of Vodafone and AirTouch," p. 30.
16. Interview with Chris Gent, June 21, 2000; interview with Arun Sarin, August 3, 2001.
17. Oliver August, "Man with the Midas AirTouch," nd, np, AirTouch Archives.
18. "Look No Wires," *The Economist,* January 23, 1999, p. 60.
19. Interview with Sam Ginn, May 1, 2001.
20. "Vodafone and AirTouch to Merge, Will Create Global Wireless Powerhouse," Press Release, AirTouch Communications, January 15, 1999.
21. Deborah Solomon, "AirTouch Still After Deal with Bell Atlantic," *San Francisco Chronicle,* January 18, 1999.

22. Peter Curwen, "Mobility Rules? Impacts of the Vodafone AirTouch–Mannesmann Take-Over," *Info* 2, no. 1 (2000), pp. 25–40. Scott Moritz, "Bell Atlantic Ponders Its Next Wireless Move," *Knight-Ridder/Tribune Business News,* January 18, 1999.

CHAPTER TWENTY-THREE

1. William Boston, "With AirTouch Deal, Vodafone Sheds Past as a Mere Upstart," *Wall Street Journal Europe,* January 19, 1999, p. A1.
2. Interview with Chris Gent, June 21, 2000.
3. Interview with Sam Ginn, May 1, 2001.
4. Andy Dornan, "Brave Old World," *Red Herring,* April 2000, p. 170.
5. Andy Dornan, "Paying a Call," *Red Herring,* April 2000, p. 178.
6. Roger Crowe, "Ginn Gets a $450m Tonic," *The Guardian,* January 19, 1999, p. 21.
7. Interview with Sam Ginn, June 18, 2001. Interview with April Walden, June 18, 2001.
8. Interview with Chris Gent, June 21, 2000.
9. "Sam-E-Gram," September 1, 1998, AirTouch Archives.
10. Interview with Amy Damianakes, April 28, 2000. Interview with Kathy Reinhart, June 18, 2001.
11. "The Live Wire of Wireless," *Business Week,* June 12, 2000, p. 80.
12. Interview with Craig McCaw, June 8, 2000.
13. Interview with Michael Miron, May 5, 2000. Interview with Amy Damianakes, April 28, 2000.
14. Interview with Michael Boskin, July 12, 2000.
15. "Sam-E-Gram," May 3, 1999, AirTouch Archives.
16. As of September 30, 1999, Bell Atlantic had approximately 7.3 million wireless customers (including its share of partnerships like PCS PrimeCo), compared to Air-Touch's 9.1 million. The GTE merger would add another 6.9 million customers, including 1.7 million that GTE acquired in businesses purchased from Ameritech in October 1999. Vodafone AirTouch, "Listing Particulars Relating to the Issue of Ordinary Shares in Vodafone AirTouch Plc in Connection with the Offer for Mannesmann AG," December 20, 1999, pp. 138, 162, 206. Interview with Chris Gent, June 21, 2000. Interview with Sam Ginn, May 1, 2001.
17. Interview with Mohan Gyani, September 20, 2000. Interview with Sam Ginn, June 18, 2001.
18. Interview with Sam Ginn, May 1, 2001.
19. Nicole Harris, "Why VoiceStream Wireless is Getting a Call from Japan," *Wall Street Journal,* May 31, 2000, p. B4.
20. Federal Communications Commission, *Second Annual Report and Analysis of Competitive Market Conditions with Respect to Commercial Mobile Services,* FCC 97-75, March 25, 1997, pp. 27–8.
21. Interview with Terry Kramer, July 11, 2000; interview with Chris Gent, June 21, 2000.
22. Interview with Chris Gent, June 21, 2000.
23. Interview with Mohan Gyani, September 8, 2000.
24. Interview with Chris Gent, June 21, 2000.
25. Deborah Solomon, "Vodafone Joins Bell Atlantic in Huge Cell Deal," *San Francisco Chronicle,* September 22, 1999.
26. "Verizon Wireless Facts at a Glance," Verizon Wireless website, June 12, 2000. Also Press Release, "Bell Atlantic and Vodafone AirTouch to Form New U.S. National Wireless Competitor," September 21, 1999.

27. Standard & Poor's, "Telecommunications: Wireless," *Industry Surveys* (New York, June 22, 2000), p. 1.
28. Peter Curwen, "Mobility Rules? Impacts of the Vodafone AirTouch–Mannesmann Take-Over," *Info* 2, no. 1 (February 2000), pp. 25–40.
29. Seth Schiesel, "Wireless Unit Spinoff Called Near at Sprint," *New York Times,* May 19, 1998, p. D3.
30. Daniel Roth, "Is Nextel the Next AirTouch?" *Fortune,* February 15, 1999, p. 195.
31. Nicole Harris, "SBC, BellSouth Confirm Venture," *Wall Street Journal,* April 6, 2000, p. B10.
32. Solomon, "Vodafone Joins Bell Atlantic in Huge Cell Deal."
33. Ironically, AT&T had experienced similar difficulties in trying to induce Pacific Telephone employees to move around in the Bell System prior to the breakup. This unwillingness to move had contributed to Pacific Telephone's relative cultural isolation within the System. Eric John Abrahamson with Marjorie Wilkens, *Learning to Compete: A History of Pacific Telesis Group* (San Francisco, 1994). SBC Communications Archives.
34. AT&T Wireless tracking stock began trading on April 27 as the market for telecom stocks was beginning to fall. The company issued 15.5 percent of the total shares and raised $10.6 billion in cash. AT&T reserved $7 billion from this sale for wireless expansion, which included buying surplus licenses from Verizon in California and Texas. "AT&T Wireless Gains 8-1/2% on Debut in Stormy Market," *Financial Times,* April 28, 2000, p. 1. Nicole Harris, "AT&T Agrees to Acquire Wireless Assets in California and Texas for $3.3 Billion," *Wall Street Journal,* June 20, 2000, p. A3.

CHAPTER TWENTY-FOUR

1. Peter Curwen, "Mobility Rules? Impacts of the Vodafone AirTouch–Mannesmann Take-Over," *Info* 2, no. 1 (2000), pp. 25–40.
2. Klaus Esser credited PacTel–AirTouch for helping Mannesmann get its start in telecommunications. "We profited significantly from AirTouch's know-how, especially during the early years of D2," he told *Global Telecoms Business* in December 1998. "And we are very happy to continue with AirTouch in this success." "Mannesmann Fixes Its Sights on Deutsche Telekom," *Global Telecoms Business* (December 1998).
3. Joachim Funk to Sam Ginn, January 18, 1999, Ginn Papers, AirTouch Archives. In a follow-up letter to Ginn, Funk expressed his concern that Vodafone–AirTouch would seek to optimize the value of E-Plus in order to boost the sale price. This would have an impact on pending struggles over various regulatory issues. See Joachim Funk to Sam Ginn, January 26, 1999, Ginn Papers, AirTouch Archives.
4. Sam Ginn to Joachim Funk, January 25, 1999, Ginn Papers, AirTouch Archives.
5. Curwen, "Mobility Rules?" Anita Raghavan, "Mannesmann's Pricing Game on Hostile Bid Irks Holders," *Wall Street Journal,* January 25, 2000, p. C1; Steven Lipin, "Union Will Shower Many with Riches," *Wall Street Journal,* February 4, 2000, p. C1. Interview with Sam Ginn, May 1, 2001.
6. Curwen, "Mobility Rules?"
7. "Mannesmann Rejects Offer from Vodafone," *San Francisco Chronicle,* November 15, 1999. See also Vodafone AirTouch, "Listing Particulars Relating to the Issue of Ordinary Shares in Vodafone AirTouch Plc in Connection with the Offer for Mannesmann AG," December 20, 1999.

8. Interview with Chris Gent, June 21, 2000.

9. Ibid.

10. Ibid.

11. Curwen, "Mobility Rules?"

12. Ibid.

13. "Chris Gent is Already Looking beyond Mannesmann," *Business Week,* February 7, 2000, p. 18.

14. Malcolm Spicer, "Vodafone Will Talk If Mannesmann Listens," *Wireless Insider,* January 24, 2000.

15. Curwen, "Mobility Rules?"

16. "Chris Gent is Already Looking beyond Mannesmann," p. 18.

17. As Peter Curwen points out, Schroder could not push the issue too stridently because the German champion, Mannesmann, had already invaded the U.K. market by purchasing Orange. And Tony Blair was quick to remind Schroder that Europe was supposed to be a single market. Curwen, "Mobility Rules?"

18. Interview with Sam Ginn, May 1, 2001.

19. "Vodafone and Mannesmann," *The Economist,* December 4, 1999, p. 62.

20. Ibid.

21. At one point, however, Gent nearly made a critical misstep. In an effort to lighten the tension over the hostile bid he poked fun at Klaus Esser's accent and dry personality. When his actions were reported in the press they angered many of Mannesmann's German shareholders. "Vodafone's Folly," *The Economist,* July 15, 2000, p. 20.

22. Curwen, "Mobility Rules?"

23. "Chris Gent is Already Looking beyond Mannesmann."

24. Institutional Shareholder Services and Lehman Brothers offered this advice in mid-January. See Spicer, "Vodafone Will Talk If Mannesmann Listens."

25. Lipin, "Union Will Shower Many with Riches."

26. Dan Sabbagh, "Vodafone Gives Gent £10m Cash and Share Bonus," *Daily Telegraph,* June 20, 2000, p. 23.

27. Curwen, "Mobility Rules?"

28. Under the terms of the deal, Vodafone AirTouch gained a 10-percent stake in France Telecom (to be liquidated when "New Orange" went public). France Telecom agreed to combine its wireless operations with Orange and make Orange's CEO, Hans Snook, the leader of the combined entity. With operations in Britain, France, Belgium, Denmark, and Eastern Europe, Orange became the second-largest wireless operator in Europe and Vodafone AirTouch's leading competitor. France Telecom announced that it planned an IPO of Orange shares, to be dubbed New Orange, soon. Snook also told reporters that, with the $13 billion he hoped to raise with the IPO, he wanted to enter the U.S. market. Gautam Naik and Kevin J. Delaney, "France Telecom Approves Deal to Buy Orange from Vodafone," *Wall Street Journal,* May 30, 2000, p. A16. Gautam Naik and Anita Raghavan, "France Telecom Confirms Plan to Buy Vodafone Unit," *Wall Street Journal,* May 31, 2000, p. A21.

29. Naik and Raghavan, "France Telecom Confirms Plan to Buy Vodafone Unit."

30. Vodafone Group Plc, "Annual Report & Accounts," 1999.

31. In the language of the industry, Vodafone AirTouch and Mannesmann's combined wireless operations as of December 1999 covered 42 million proportionate customers and had the potential to serve 512 million proportionate POPs. Vodafone

AirTouch, "Listing Particulars Relating to the Issue of Ordinary Shares in Vodafone AirTouch Plc in Connection with the Offer for Mannesmann AG," December
20, 1999.

32. Interview with Chris Gent, June 21, 2000. Also, interview with Michael Boskin, July
12, 2000.

33. Vodafone AirTouch Plc, "Annual Review and Summary Financial Statement,"
March 2000, p. 27.

34. "Vodafone's Folly."

35. Ibid.

36. Carlta Vitzthum, "Vodafone Offer for Airtel Stake is Likely to Prevail," *Wall Street
Journal,* December 27, 1999, p. A15. Keith Johnson, "Vodafone Wins Spanish Foothold with Deal for Control of Airtel," *Wall Street Journal,* June 13, 2000, p. A21.

37. Paul Betts, "Rising Star in the Mobile Galaxy," *Financial Times,* April 28, 2000, p.
16.

38. Vodafone Group Plc, "Interim Report," Mar.–Sep. 2000, p. 3.

39. "Vodafone to Acquire Eircom's Eircell," *Wall Street Journal,* December 22, 2000,
p. A11.

40. Gautam Naik, "Vodafone Now Looks toward Asia," *Wall Street Journal,* November 15, 2000, p. A21.

41. Henny Sender, "Vodafone Move Highlights Growing Allure of China," *Wall Street
Journal,* October 6, 2000, p. A21. Vodafone Group Plc, "Interim Report," p. 12.

42. Anita Raghavan and Gautam Naik, "Vodafone Set to Finalize Two Big Deals," *Wall
Street Journal,* December 20, 2000, p. A14.

43. Anita Raghavan and Nikhil Deogun, "Vodafone Set to Expand Japan Stake," *Wall
Street Journal,* February 26, 2001, p. A14.

44. "Vodafone's $5 Billion Offering Sets Europe Record," *Wall Street Journal,* May 3,
2001, p. A12.

45. "Telecoms in Trouble," *The Economist,* December 16, 2000, pp. 77–9.

46. "Vodafone's $5 Billion Offering Sets Europe Record."

47. Nancy Gohring, "Ericsson/Qualcomm Bitter Feud Ends," *Telephony,* March 29,
1999, p. 8.

48. Henrik Glimstedt, "Competitive Dynamics of Technological Standardization: The
Case of Third Generation Cellular Communications," *Industry and Innovation* 8,
no. 1 (2001), pp. 49–78.

49. George Gilder, *Telecosm: How Infinite Bandwidth Will Revolutionize Our World*
(New York, 2000), p. 92.

50. "The Really Pervasive, Really Fat Wireless World," *Investor Entrepreneur Network,*
⟨www.siliconindia.com⟩.

51. Janet Guyon, "What Does This Gent Really Want?" *Fortune,* March 6, 2000, p.
163.

52. "Vodafone, Vivendi Plan Major Internet Alliance," *Wall Street Journal,* January 31,
2000, p. A21. Curwen, "Mobility Rules?" Vodafone Group Plc, "Interim Report,"
p. 11.

53. Guyon, "What Does This Gent Really Want?"

54. "Mobile Licenses," *Financial Times,* April 28, 2000, p. 20.

55. Nancy Gohring, "Up in the Air," *Interactive Week,* January 9, 2001.

56. "Telecoms in Trouble," p. 77.

57. "Vodafone's $5 Billion Offering Sets Europe Record."

CHAPTER TWENTY-FIVE

1. "Sydney Welcomes World with Opening Ceremonies," CNN Sports Illustrated (CNNSI.com), September 16, 2000. "Dialing for Gold," *Wall Street Journal,* September 18, 2000, p. A6.

2. Jennifer Tomaro, "Telstra's Gold-Medal Olympic Communications Infrastructure," CommWeb.com, September 27, 2000.

3. After selling his company, however, McCaw diversified his investments in telecommunications, making major investments in both wireless and wireline systems. Some speculated that he was working to build his own Baby Bell from scratch. O. Casey Corr, *Money from Thin Air* (New York, 2000).

4. Richard R. Nelson, "The Co-evolution of Technology, Industrial Structure, and Supporting Institutions," in Giovanni Dosi et al. (Eds.), *Technology, Organization, and Competitiveness: Perspectives on Industrial and Corporate Change* (Oxford, 1998), pp. 319–37.

5. Alfred D. Chandler, Jr., *Scale and Scope: The Dynamics of Industrial Capitalism* (Cambridge, MA, 1990).

6. Alfred E. Kahn, *The Economics of Regulation: Principles and Institutions,* 2 vols. (Cambridge, MA, 1988).

7. Martha Derthick and Paul Quirk, *The Politics of Deregulation* (Washington, DC, 1986). Richard H. K. Vietor, *Contrived Competition: Regulation and Deregulation in America* (Cambridge, MA, 1994).

8. Nathan Rosenberg, *Exploring the Black Box: Technology, Economics and History* (Cambridge, 1994), p. 15.

9. Interview with Sam Ginn, July 11, 2001.

10. As Arun Sarin points out, even with this new improved version of CDMA, "We did not get a global standard. There are five versions of WCDMA, none of which talk to each other." Interview with Arun Sarin, August 3, 2001. To that extent, technological globalization in wireless is far from complete.

11. Harald Gruber, "Spectrum Limits and Competition in Mobile Markets: The Role of License Fees," *Telecommunications Policy* 25 (2001), pp. 59–70.

12. Nelson points out that this idea was first suggested by Schumpeter and Veblen and amplified recently by Carlota Perez, "Structural Change and the Assimilation of New Technology in the Economic and Social System," *Futures* 15, no. 4 (1983), pp. 357–75, and C. Freeman, "The Nature of Innovation and the Evolution of the Productive System," in *Technology and Productivity* (Paris, 1991), both cited in Nelson, "The Co-evolution of Technology, Industrial Structure, and Supporting Institutions," pp. 319–37.

13. John H. Dunning, "Governments and the Macro-Organization of Economic Activity: A Historical and Spatial Perspective," in John H. Dunning (Ed.), *Governments, Globalization, and International Business* (New York, 1997), pp. 32–7. See also William Lazonick, "Business Organizational and Competitive Advantage: Capitalist Transformations in the Twentieth Century," in Giovanni Dosi et al. (Eds.), *Technology and Enterprise in Historical Perspective* (Oxford, 1992).

14. For an interesting description of the effect of stock options on the psychology of employee compensation in Silicon Valley, see Michael Lewis, *The New New Thing: A Silicon Valley Story* (New York, 2000), p. 86.

Index

Index

Index

Index

Index

Index

Index

Index

Index

Index